D0163242

Contact with the stars

'Is there intelligent life on Earth?'
'Yes, but I'm only passing through'

—Graffito in the Mathematics Department,
University of Cambridge

Contact
with
the stars

The search for extraterrestrial life

Reinhard Breuer

Translated from the German by
Cecilia Payne-Gaposchkin
and Mark Lowery

W. H. Freeman and Company Limited
Oxford and San Francisco

W. H. Freeman & Company Limited
20 Beaumont Street, Oxford, OX1 2NQ
660 Market Street, San Francisco, California 94104

Library of Congress Cataloging in Publication Data

Breuer, Reinhard A., 1946–
    Contact with the stars.

    Translation of: Kontakt mit den Sternen.
    Bibliography: p.
    Includes index.
    1. Life on other planets.   2. Interstellar communica-
tion.   I. Title.
QB54. B69813   999   81–9908
ISBN 0–7167–1355–1   AACR2
ISBN 0–7167–1389–6 (pbk)

Original German language edition © Umschau Verlag Breidenstein KG,
Frankfurt am Main, 1978

This translation © W. H. Freeman & Company Limited, Oxford, 1982

No part of this book may be reproduced by any mechanical,
photographic, or electronic process, or in the form of a
phonographic recording, nor may it be stored in a retrieval
system, transmitted, or otherwise copied for public or
private use without the written permission of W. H. Freeman
and Company Limited

Setting by Macmillan India Limited
Printed in the United States of America

QB
54
B69813
.1982

Contents

part III
The conquest of space   233

Preface to the German edition

To write a book in 1978 about extraterrestrial intelligence and the chances of communication with other civilizations in our Galaxy is not only topical, in view of the current preoccupation with the subject, but also seems to me to be particularly necessary. It is only 18 years since a telescope in Green Bank, West Virginia, was turned to another star for the first time, to listen for artificial radio signals. For mankind, this was the beginning of a *conscious* search for interstellar communication. Unintentionally, the Earth has been transmitting since radio technology began: for around 50 years, artificial radio signals have been streaking through space at the speed of light, reaching some 20 new stars every year. If, on some planet, perhaps 50 light-years away, an intelligent being were able to direct a radio receiver at the Earth, interstellar contact between two civilizations of the Milky Way would have been established. *Contact* is taken in this book to mean not only two-way communication but also the one-way reception of a message from a second party (or simply the discovery of an archaeological remnant of another civilization). Unilateral reception of radio signals is the most probable form of contact in a 'communications network' in which a signal may travel for 30 000 years before reaching a receiver.

One further semantic clarification before we go on: occasionally in this book we will speak of the 'optimistic' standpoint. This implies among other things the supposition that our home Galaxy is populated by millions of technological civilizations; that these include numerous highly developed civilizations which have already been in contact with one another for millions of years, as members, so to speak, of a 'Galactic Club'; that these have long since overcome the dangers of cultural self-annihilation, developed the art of biochemical gene synthesis beyond the complexity of human genetics, and mastered laws of physics which are still a closed book to us; and that they can draw on the energy of stars and use black holes to travel through space just as we would use underground railways.

I consider this book to be necessary because the time has come to take a critical look at these still-prevalent optimistic views regarding life on other stars, in the light of recent findings in various branches of science: in interstellar chemistry with respect to chemical evolution in interstellar gas and on comets and meteoroids; in biophysics and molecular biology as regards the processes which mark the actual beginning of life; in astrophysics concerning the origin of our Solar System and the development of the Earth's surface as the arena for the rise of our civilization; in planetary research, with its data on life on Venus, Mars and Jupiter; and not least in radio, X-ray and even gravitational-wave astronomy for the *evidence* these can provide for highly developed extraterrestrial civilizations.

It is only now that the *necessary correction* to these over-optimistic views, which we derive from these findings, has become possible; for the research results which seriously question the tenets of the optimistic standpoint date only from

1977. I have tried to present these most recent results, which are in part of revolutionary significance for the question of extraterrestrial life, in a reasonably compact way, and to confront them with what has hitherto been thought about extraterrestrial intelligence. In this, I have taken note not only of the suggestions of scientists but also on occasion of ideas and hypotheses taken from science fiction. My aim was to determine, by the end of this confrontation, what may in my opinion be taken as factually refuted, what must remain unverifiable wild speculation, and what may be retained as an interesting hypothetical possibility or a more or less plausible scientific thesis. A summary of the main arguments for my 'pessimistic' standpoint is given at the end of Part II, in the section 'We are (still) alone' (page 224).

Without the help of a number of friends and colleagues, this book could not have been completed in its present form. Above all, I am indebted to Dr W. Ochs, Privatdozent at the University of Munich, for the suggestions, discussion and criticism he contributed to the development of the manuscript. I am grateful to P. Kafka (Max Planck Institute for Physics and Astrophysics, Munich) and to Professor M. Eigen and Dr B. Küppers (both of the Max Planck Institute for Biophysical Chemistry, Göttingen) for giving freely of their time. Helpful suggestions were also received from Professor G. Ludwig (University of Alberta, Edmonton) and Dr H. von Voithenberg. Particularly warm thanks are due to Frau Anneliese and Dr Josef Zintel, who provided me with accommodation during the writing of the text.

Munich, February 1978   Reinhard Breuer

Preface to the English edition

Since this book was completed, a host of new discoveries—particularly about the Solar System—have been rained upon us. The most important results came from the NASA space probes Voyager 1 and 2, which revealed many astonishing facts about Jupiter and its moons and from the rings of Saturn. Among them were the discovery of the first active volcanoes outside Earth on one of the Galilean moons, Io, powered by the strong tidal forces of Jupiter; the unsuspected rings around the planet; a fourteenth moon inside the hitherto closest-known moon, Amalthea; active thunderstorms with lightnings in Jupiter's atmosphere; and the properties of Jupiter's Red Spot, a gigantic whirlpool whose red colour is most likely caused by phosphorus. Continuing its interplanetary journey, Voyager 1 sent a series of close-ups of Saturn's rings in November 1980. The main hope for evidence of organic life had centred on the moon Titan, but Titan was found to be hidden under layers of haze in its upper atmosphere, containing only trace amounts of organic molecules produced at -180 degrees Celsius by photo-chemical reactions.

Another host of new discoveries concerns Venus, diligently scrutinized by an armada of space probes since December 1978. Despite all these impressive results, no positive evidence has been produced for any kind of life within the Solar System. The organic molecules spotted in Jupiter's atmosphere, on the other hand, support the findings made in observing interstellar gas—that the fundamental building-blocks of life can come into being under a very wide range of circumstances. However, the search for extraterrestrial artificial signals has continued completely without success.

Thus, the main conclusions of the book—that some sort of primitive life may be quite common, but technological civilizations are likely to be extremely rare within our Galaxy (see my summary of the book at the end of Part II, page 224)—are unchanged by the latest developments.

On the communication side, an intriguing new possibility was suggested in Spring 1979 by two Japanese scientists: the biological channel. It is conceivable that interstellar messages are encoded in the DNA of microorganisms which have been sent to other planets, such as Earth. (This method offers a number of advantages.) The organism analysed, the bacteriophage $\Phi X 174$, did not appear, however, to contain a message.

One most surprising development has grown in importance since this book appeared in German. The topic of contact with extraterrestrial life-forms has received increasing attention, while at the same time—as first emphasized in this book—all the evidence points more and more against the presence of tech-nological civilizations or supercivilizations. While the Universe appears to be more complex and fascinating with each new discovery made, it seems to be devoid of a Galactic Club of life-forms supervising us, or even willing to

communicate with us. We may be quite alone—the investigation of how this has
come about now constitutes a major challenge to terrestrial science.

Munich, November 1980   Reinhard Breuer

Contact with the stars

part I
From molecule to man

The unknowable birth of the Universe

Once, the Universe was seen as a temple: on the disc of the Earth rested the 'pillars of the firmament'; on these pillars, like a roof, the firmament itself. From hatches in the firmament fell the 'sweet waters of heaven', for the Lord divided the waters, the sweet from the salty. And finally, attached to the firmament were the planets and fixed stars. This concept of the Universe was the basis for the cosmology of the Christian West; it is derived from the Bible and appears little changed in ancient Greece, where, with all the refinement of a complex mathematical theory, the rigid firmament was replaced by dozens of crystal spheres rotating in different directions. It was only within the framework of this model, in which the world was arranged like a house, that the human race could find refuge, an 'inside' and an 'outside'; and with this arose the questions: What is 'outside'? What existed 'before'?

Modern cosmology, which began with Albert Einstein's *General Theory of Relativity* and Alexander Friedman's model of the Universe, has departed radically from this down-to-earth cosmic dwelling for humanity. The discovery that the Universe 'began' in very hot and dense conditions many millions of years ago (by current theory, about 18 000 million—18^9—years ago) robbed mankind of its infinite past, in which even Einstein had originally believed. Using modern methods, we can calculate back as far as an original hot and dense state of the Universe; of matters before that, theory can tell us nothing. Whether we choose to call this phase 'the beginning' is a question of semantics.

The revolution in thought was even more drastic. Einstein showed that Newton's absolute, God-given space did not exist either. Einstein united the previously separate concepts of absolute space, enduring time and immutable matter: at the birth of the Universe, the so-called 'Big Bang', there came into being a structure of space, time and matter, inseparably combined, and with the passage of time space itself altered. At the commencement of time we do not have—as in the common misconception—a fireball exploding like a bomb into pre-existing but still empty space; rather, this original explosion filled from the start all of space, which was itself expanding. Already the demands on our normal powers of imagination are too great: provided with a language and way of thinking in which spatial and temporal concepts are deeply embedded, we cannot conceive of the absence of space and time. However, we are not yet totally helpless: on the one hand, we can settle how far back physics can provide answers (and where it leaves gaps), and on the other, we can recognize the meaninglessness of questions which arise entirely from the semantic difficulty we have mentioned. For, although the cosmic world-picture has changed radically, human language, whose content often bears the stamp of a naive cosmology, has not. Thus, it is often asked, as though the world were still a Babylonian temple: 'What existed *before* the Big Bang? What is *outside* space? What is the origin of the Universe?' The first two questions are meaningless in the context of the General Theory of Relativity, as the words 'before' and 'outside' assume time and space, which only came into being as a result of the Big Bang itself.

What, then, are the 'final' meaningful statements we can make about the Big Bang; how far back will the physical theory of the Universe take us? To answer these questions, we can only work on the basis of the General Theory of Relativity and the other laws of physics valid on Earth; the assumption is that they also applied at the time of the Big Bang. Einstein's theory, for instance, has only been properly tested within the Solar System, that is, only in very weak gravitational fields. Besides this, it embodies no quantum structure, and will probably fail when we reach back as far as the so-called Planck time, 10^{-43} seconds after zero time. For the time before this, our calculations must take Quantum Theory into account, and as a result it is no longer possible to trace historical events with any certainty. Thus, at around the Planck time there is a barrier which in all probability makes it impossible for us to reconstruct the original conditions.

However, our present knowledge of the past state of matter stops far short of this obstacle. Mixtures of matter and radiation in conditions of extreme density and temperature can give rise to processes which, with even our most advanced particle accelerators, we cannot imitate and thereby examine. Although these laboratory experiments have provided high-energy physicists with (very vague) information on the properties of matter up to the density of atomic nuclei ($10^{13} \times$ the density of water), this concerns very *cold* matter. It is only when the temperature of the original matter has fallen below 100 000 million degrees Celsius ($10^{11}\,^\circ$C)—about one-hundredth of a second after the Big Bang—that even high-energy thermodynamics can make any firm statements.

Just where—looking backwards—high-energy physics itself drops out of the running, is something which offers scope for speculation. What was the initial state of the Universe like? It is possible that there never was an infinitely dense initial state, but only, at the transition from a preceding contraction phase, a temporary condition of measurable heat and density—a stage in the life of an oscillating universe. Admittedly, the General Theory of Relativity in its pure, classical form would contradict this, but the validity of Einstein's theory is not necessarily universal. Up to now, however, no traces of a previous universe have been discovered: radioactive elements have invariably proved to be younger than the period since the Universe began to expand. If they had originated in the preceding phase, they would naturally be older. We must remember, though, that with temperatures above 100 000 million $^\circ$C and with space itself reduced to only a few light-years in extent during the transition from contraction to expansion, atomic nuclei would in any case have been broken down into their elementary particles by gamma radiation, and all information from the earlier phases would have been wiped out. (A light-year is the distance travelled by a light ray in a year, that is, 9.4505×10^{15} metres.)

Throughout the whole of the first second, a state of equilibrium prevails, in which every variety of matter—principally short-lived elementary particles—converts to radiation, and the energy-rich radiation in turn produces matter. Overall, radiation energy dominates cosmic evolution in this phase; matter still plays a negligible role, remaining for the moment an insignificant 'trace effect'. But by the end of the first second the advancing process of cooling allows the first kind of particle to condense out: the massless neutrinos and their anti-particles. These decouple from the equilibrium state and spread throughout space unhindered by matter; from this point on, the Universe is largely transparent for neutrinos. With suitable neutrino telescopes, we could theoretically pick up these particles and use them to see at least as far back as the end of the first second of creation. The gravitational waves produced by the Big Bang are all that are likely to derive from a still earlier phase. But detectors for these fantastically weak gravitational waves are still under construction, and must be made considerably more sensitive if they are to prove their existence successfully (see Part II: 'The curious waves of gravity', page 179).

Only after the passage of the first 13 seconds does the temperature of the Universe fall to 3000 million °C—cool enough, as Stephen Weinberg explains in detail in his book *The First Three Minutes*, to permit the first particles of matter to 'freeze out'.[1]

Now electrons, the particles of the atomic shell bearing a negative electrical charge, and their anti-particles, the positrons, come into being, capable of withstanding the intense radiation. The building blocks of atomic nuclei, protons and neutrons, also appear at this time. With protons, material for the nuclei of the later hydrogen atoms is available; and a combining of protons and neutrons—nuclear fusion—brings into being nuclei of 'heavy hydrogen' or deuterium: nuclei with one proton and one neutron each. Along with these, the first helium nuclei are being formed, with two protons and two neutrons each (helium-4). Production of helium nuclei continues for some time, coming to a halt only when the temperature falls below 1000 million °C (to be exact, at 900 million °C). The Cosmos is now 3 minutes and 40 seconds old; at this point, nearly all remaining neutrons have been used up in the production of helium nuclei, which now make up 29 per cent of matter.

As the Universe expands further, the temperature falls to a point where the available atomic nuclei combine with the electrons to form complete neutral hydrogen and helium atoms. Photons (the smallest units of energy in electromagnetic fields), which up to now have been repeatedly scattered by free electrons, are suddenly able to decouple from matter and penetrate it unhindered; the Universe now becomes as 'transparent' to photons as it already is to neutrinos and gravitational waves. At this point the equilibrium hitherto maintained between radiation and matter is at last destroyed, so that the dominating role in the Universe, which previously fell to radiation, is taken over by matter. According to the current interpretation, this transition took place in the year 700 000 after zero. The cosmic gas, at that time, had cooled to 3500 °C. Only after this could the long-distance effects of gravity magnify the irregularities and 'lumps' in the distribution of matter in such a way that galaxies, stars and eventually planets could come into being.

Apart from hydrogen and helium, none of the higher chemical elements yet exists. The physical conditions between the Big Bang and the final decoupling do not even allow the synthesis of the next element in the build-up, carbon-12. Why did the Big Bang synthesis progress only as far as helium? Theoretically, two helium nuclei can combine to produce first beryllium-8 and then, by the addition of a further helium nucleus, carbon-12. But beryllium-8 is radioactive, and decays too quickly to be available for the second step in this synthesis: the carbon nuclei, therefore, would have had to be formed directly, by the collision and fusion of three helium-4 nuclei. But by the time when there first were enough helium nuclei to make possible a synthesis of carbon, the expansion of the Universe had reduced the density of the particles too greatly. Collisions of three helium nuclei thus occurred far too rarely, and at the end of the hot phase following the Big Bang, helium remained the heaviest element.

In the next 18 000 million (18×10^9) years, the electromagnetic radiation which has filled the Universe since its decoupling, practically undisturbed by matter, cools further and loses its energy as a result of the continuing expansion of the Cosmos. In fact, we can now detect this 'echo' of the final stage of creation as 'cosmic background radiation': it represents the present limit of astronomical observation into the past. This residual radiation was discovered accidentally by the American scientists A. A. Penzias and R. W. Wilson in Holmdel, New Jersey, in 1965. Their actual intention had been to look for radio waves from the Galaxy with a 6.5-metre horn-shaped aerial of the Bell Telephone Company. It later transpired that a team led by R. H. Dicke at Princeton University had been about to start searching for this background radiation: they had developed a theory of a hot Big Bang which predicted its existence. After its accidental discovery it also turned out that the Russian-American physicist George Gamow had made a similar prediction as early as 1948. On the basis of theory, it was expected to have the type of radiation spectrum, wavelength for wavelength, characteristic of a black body at 3 degrees Kelvin. By means of measurements made by balloon in 1975, earlier data were shown to be incorrect, and the black-body characteristics were also confirmed for a hitherto inaccessible range of wavelengths. This cosmic background noise reaches us in equal strength from all directions in the form of radio waves between 2 millimetres and 10 centimetres in wavelength. This regularity in its angular distribution, known as *isotropy*, shows that at the time of uncoupling the Universe must already have been well organized—a surprise for cosmologists, who were more inclined to assume a chaotic initial condition. However, the hypothesis regarding the isotropy of space has now been tested to an accuracy of one part in a thousand by examining the angle dependence; so any original chaos which had existed must have damped itself out remarkably quickly.

Less is known about the future development of the Universe. If expansion continues for all time, we would have an open Universe; if, on the other hand, the expansion stops at a maximum radius and gives way to contraction—just as a ball thrown into the air falls back again—we would have what is known as a closed Universe.

By observing the average density of cosmic material, it can be established which of these is the case. (Information on this is also provided by the speed of recession of galaxies.) Only if this density were to exceed a certain value would the Universe be closed, in which case, in the distant future, the effects of gravitation would cause it to fall in on itself again. In 1973 the Americans Allan Sandage and Eduardo Hardy (Pasadena, California) announced a value for the density which suggested a closed Universe; however, this situation was soon upset by considerations of the internal development of galaxies, which are especially relevant for the more distant (and therefore younger) galaxies, as well as by more recently developed methods of observation and evaluation of the data. Using the analyses of galactic development pioneered by Beatrice Tinsley (Austin, Texas), James Gunn and J. B. Oke (Hale Observatory) have established values for the density which would be appropriate to an open Universe. But these results should not be overvalued: the large error margins typical for such observations preclude any firm conclusions, as does the fact that Tinsley's assumptions on the age and development of galaxies are as yet only theory, which cannot be tested by observation. However, there are numerous indications that the density of matter is no more than one-tenth of the critical value necessary for a closed Universe. The occasional discovery of intergalactic gas clouds does not alter this: this matter, invisible except during its occasional appearance in the light of quasars, is still too poorly understood for us to draw from it any generalized conclusions as to the open or closed nature of the Universe—though this is a game which science journalists are unfortunately only too ready to play.

The Big Bang model explains many facts, but nevertheless it has its shortcomings. For instance, measurements show that material particles and photons occur in a proportion of 1 to about 300 million in space. So far, attempts to derive this empirically established relationship theoretically from the model have not been wholly successful.

A related question is: Where did all the anti-matter go? For if, in the first ten-thousandth of a second after the Big Bang, all matter came into being by what is known as pair production, it might be expected that there should be just as many anti-particles as particles. Despite this, we find ourselves surrounded exclusively by 'normal' matter; of anti-matter there is no sign. And whether perhaps other galaxies are composed entirely out of anti-matter, which is at least conceivable, is a point which can only be settled indirectly and with difficulty.

The current thinking is that shortly after the Big Bang there may have been slightly more matter around than anti-matter. The reason for this may be found in processes violating the particle (baryon) number, as suggested by recent 'Grand Unified Theories'.

Nor are the details of the birth of galaxies yet understood within the framework of the Big Bang model. There are certainly a large number of interesting suggestions, but none has yet been able to gain full acceptance. The dilemma is a simple one: *within* the early phase, which was dominated by radiation, 'bunching' of matter was prevented. Matter was very evenly distributed, as shown by the isotropic nature of the cosmic background radiation now observed. *After* the decoupling of radiation and matter, it was not possible for the very small irregularities then present in the distribution of the gas to increase quickly enough to form entire galaxies. Possibly there were very large black holes which, in the course of thousands of millions of years, attracted through their gravitational fields matter of sufficient mass to form whole galaxies. These problems are also connected with the question of the uniqueness of the Universe: why is the Universe as it is—could it have turned out differently? Natural laws only allow conclusions of the nature, 'if A . . . then B'; they tell us nothing about the 'if ', the so-called initial conditions. One goal of cosmologists is to develop a model for the Big Bang in which the many mathematical possibilities for the initial conditions would have no consequences for the present state of the Universe. This would be the case if all conceivable initial conditions—chaos or early order—led to the same eventual state of the Universe, that is, to the Universe we observe today. In other areas of physics there are examples of such behaviour. It makes practically no difference to a pendulum clock whether the pendulum is set swinging with a strong or weak push: after a time it will swing with exactly the same period of oscillation. In liquids, too, certain turbulences will always introduce themselves, independent of the precise nature of the disturbance. However, cosmology still lacks a theory to specify the right starting conditions to give the result that the Universe as we know it is the only possible one.

Of course, although we do not know the answers to these questions, galaxies have nevertheless somehow come into being. Our home galaxy, the Milky Way, is a flat disc with a dense central region (see Figure 1). With its spiral arms, it has a diameter of 100 000 light-years and a thickness of the order of 6000 light-years. Twelve thousand million (12×10^9) years ago, the gaseous mass of our Galaxy still rotated in the form of a sphere; the sphere then shrank under its own gravitation until the pressure of the heated gas and the centrifugal force created by the rotation put a stop to the contraction process. During this development, the cloud of gas flattened into a disc as a result of the conservation of angular momentum. Other galactic nebulae, which started with a slower rotation, have more or less retained their spherical shape; these are the elliptical galaxies. During the contraction, older stars remained in the earlier spherical region, the so-called halo; these old

Plan view

Sun

10 000 light-years

Profile

Sun

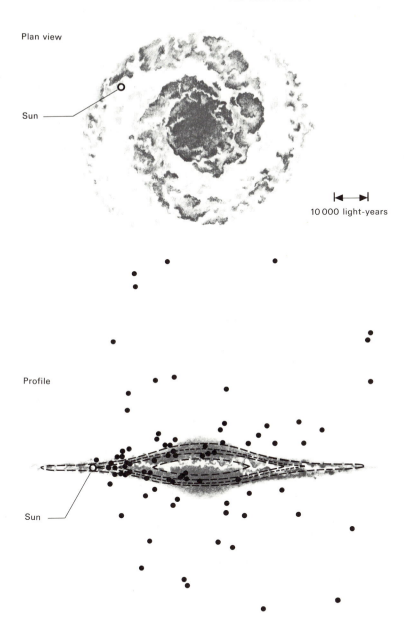

Figure 1 Our Galaxy, as seen from 'outside'.

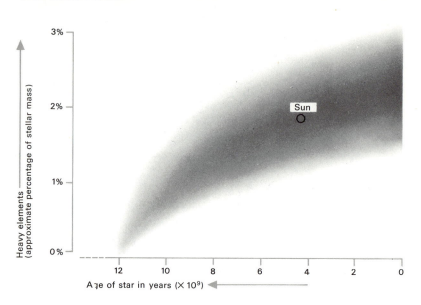

Figure 2 Heavy elements in stars of various ages. Old, dying stars exploding as novae enrich the interstellar gas with heavy elements; new stars evolve out of this material. The abundance of heavy elements is thus greatest in young stars.

stars still populate the halo and the galactic centre. Younger stars occur within the disc, with a preference for the spiral arms, most of them still associated with the gas clouds from which they were born.

A good deal can be learned from the occurrence of heavy elements in these stars. As a rule, older stars contain little of the heavy elements; but during the past 12 000 million years the proportion of heavy elements has grown to as much as 2 per cent in the youngest stars (see Figure 2). The only mechanism known to us which can produce heavy elements is nuclear fusion: protons combine to give helium nuclei, and from these carbon and the higher elements are formed. These reactions are possible hardly anywhere except in the interiors of stars; nowhere else is the temperature or the density of matter high enough. The longest process is the fusion of hydrogen to helium, which at the same time provides the lion's share of exploitable fusion energy. However, the heavier the elements involved in the fusion process, the smaller is the amount of energy gained per reaction and the sooner is each additional stage of combustion completed. The fusion chain comes to an end when the gas is no longer able to maintain a temperature sufficient to balance out the effects of its own gravity: this occurs when iron-56 is reached as an end-product. When the temperature falls, so do the gas and radiation pressures; as soon as the gas pressure diminishes, this instability may well bring about an explosion, a so-called supernova. Since the time of Christ, at least 11 supernova explosions have occurred in our Galaxy.

The Crab Nebula is the remnant of a supernova which took place in AD 1054. This was observed by Chinese astronomers of the Sung Dynasty: 'In the first year of the period Chih-ho (1054), in the fifth moon, on the day Chi-ch'on (4th July), there appeared a Guest Star . . . After more than a year it slowly became invisible.' (Quoted from ref. 2.)

In a supernova, the heavy elements already built up from hydrogen are once more dispersed into interstellar space. This brings about a continuing cycle: new stars, formed out of this enriched gas, begin their initial nebular phase already containing a greater proportion of heavy elements.

There remains a gap, however, in this scheme of the development of the higher elements: the lighter elements, especially lithium-6, beryllium-9 and the element boron with its isotopes, cannot have been produced in stellar interiors. Even at as low a temperature as $100\,000\,°C$ they would have been destroyed again. These elements, therefore, must have been produced by a 'cold' process, outside stars, and their origin is assigned to a different mechanism: it is thought that they owe their existence to bombardment by cosmic rays. According to our present understanding, cosmic rays are fired into space by the remains of supernovae. In their magnetic fields, charged particles are accelerated to close to the speed of light. When particles carrying this amount of energy encounter atomic nuclei in the interstellar gas, the nuclei are split apart: nuclear fission takes place, and the resulting fission products may include light elements. Such high-energy collisions, however, are extremely rare: for 1 gram of boron, cosmic rays must bombard 1 cubic kilometre of the interstellar medium for all of 3000 million years.

Prebiotics in interstellar space: from hydrogen to hexose

Conditions look none too promising: the temperatures are below $-250\,°C$ (that is, less than 23 degrees above absolute zero); each cubic centimetre is occupied on average by roughly three gas molecules; every hundred years a molecule captures an ultraviolet photon; and only every 3 million years does a molecule collide with a high-energy photon of cosmic radiation. In the space between the stars, chemical reactions proceed at an exaggerated snail's pace. The region seems too inhospitable to be of interest to organic chemistry. And yet, at least in terms of volume, it is the largest witches' cauldron in the Universe, with true organic reactions taking place.

It is thanks to radioastronomy that chemists are suddenly taking notice of the interstellar formations within our Galaxy—the dark nebulae and those cold clouds of gas and dust which are considered to be the precursors of new stars and planetary systems. Suddenly, there has come into being a new field within chemistry, *interstellar chemistry*; it is concerned with the chemical, 'prebiotic' evolution of interstellar molecules in a space environment.[3-7] The evolution of organic molecules was hitherto considered to be possible only in the primeval soup of a planet; but today it is increasingly clear that chemical evolution here is largely limited to special cases, and that these, moreover, arose very late in the history of the Universe: more than 10 000 million (10^{10}) years after the Big Bang. The oldest, and at the same time the largest breeding-ground of organic molecules is interstellar space.

The first molecule with more than two atoms to be discovered in space— ammonia (NH_3)—was found in 1968. The observation was made by a team led by the American Nobel prizewinner C. H. Townes, using a radio telescope on wavelengths[8] in the millimetre range—to be precise, at 1.3 centimetres.[8] The diatomic cyanogen radical (CN), the methylidyne radical (CH) and its ion (CH^+) had already been identified as components of the interstellar gas in 1940. Molecules radiate principally in the microwave region when they change their rotation state. According to quantum theory, the rotation states of a molecule are 'quantized' into a number of discrete states each with a certain energy, and a change from one state to another is accompanied by a burst of radiation corresponding to the energy difference; this energy determines the wavelength of the radiation, and gives rise to a characteristic line in the spectrum. When many transfers of this nature are taking place, for instance in the molecules of a gas cloud, whole series of spectral lines can be seen—so-called 'rotation spectra'. Each molecule has a distinctive rotation spectrum by which it can be identified— we might call it the molecule's electronic 'fingerprint'. Most of the molecules now known to be present in space—so far numbering around 50, both organic and inorganic (Table 1)—gave themselves away through the quantum jumps in the rotation of their molecular atoms and in their longitudinal vibrations. Radio-astronomers have also encountered the unknown: seven rotation spectra have been found which could not be assigned to any molecule known on Earth. This happened, for instance, with hydrogen isocyanide (HNC) in 1971. This differs from normal hydrogen cyanide (prussic acid, HCN) in having the atoms C and N transposed. In the course of their observations, American astronomers discovered a line which corresponded to no known chemical compound: its frequency was 90.66359 gigahertz. Taking up this information, researchers at the University of Giessen and the Max Planck Institute for Radioastronomy in Bonn began experiments designed to permit observation of the rotation spectrum of hydrogen isocyanide in the laboratory.[7] At first, it was difficult to produce HNC at all: conventional chemical methods failed, as the atoms of the HNC molecule usually re-arranged themselves into 'normal' prussic acid. Finally, HNC was successfully produced in gaseous form and the first rotational transition was measured. The spectral line had a frequency of 90.663 592 gigahertz—the mystery of the line was solved.

Table 1 Some of the molecules observed in interstellar space

DIATOMIC			
H_2	Hydrogen‡	$CH^\bullet$	Methylidyne radical*‡ §
$OH^\bullet$	Hydroxyl radical†‡	CH^+	Methylidyne cation‡ §
SiO	Silicon monoxide	$CN^\bullet$	Cyanogen radical*‡
SiS	Silicon sulphide§	CO	Carbon monoxide†‡
NS	Nitrogen sulphide	CS	Carbon monosulphide
SO	Sulphur monoxide		
TRIATOMIC			
H_2O	Water†	$C_2H^\bullet$	Ethynyl radical*
N_2H^+	(un-named cation)* §	HCN	Hydrogen cyanide
H_2S	Hydrogen sulphide	HNC	Hydrogen isocyanide*
SO_2	Sulphur dioxide	HCO^+	Formyl cation*
OCS	Carbonyl sulphide	$HCO^\bullet$	Formyl radical
4-ATOMIC			
NH_3	Ammonia	HNCO	Isocyanic acid
HCHO	Formaldehyde†	HCHS	Thioformaldehyde
C_2H_2	Acetylene	C_3N	Cyanoethynyl*
5-ATOMIC			
HC_3N	Cyanoacetylene§	H_2CNH	Methanimine§
HCOOH	Formic acid	H_2NCN	Cyanamide* †
6-ATOMIC			
CH_3OH	Methanol	$HCONH_2$	Formamide
CH_3CN	Methyl cyanide		
7-ATOMIC			
CH_3CHO	Acetaldehyde	CH_3NH_2	Methylamine
CH_3C_2H	Methylacetylene	CH_2CHCN	Vinyl cyanide
HC_5N	Cyanodiacetylene§		
8-ATOMIC			
$HCOOCH_3$	Methyl formate		
9-ATOMIC			
CH_3CH_2CN	Ethyl cyanide	$(CH_3)_2O$	Dimethyl ether
HC_7N	Cyanotriacetylene§	C_2H_5OH	Ethanol
11-ATOMIC			
HC_9N	Cyanotetra-acetylene§		

* Rotation spectra for this molecule were first observed in interstellar space.
† Also observed in nearer galaxies.
‡ Also observed with optical telescopes.
§ Not known on Earth.

The most complicated molecule so far may have been discovered by the British researchers Fred Hoyle and Chandra Wickramasinghe in 1977 in the Trapezium Nebula: its identification is still disputed. In their book *Lifecloud*, they suggest that the molecule is one of the so-called hexose group, to which the grape sugar glucose also belongs.[9] This molecule, $(H_2CO)_6$, has all of 24 atoms; but its identification is still very much in dispute: it could, for instance, equally well be an inorganic silicate compound. What is astonishing is that such complex molecular combinations can be produced at all in an environment as inimical to chemical reactions as an interstellar gas cloud. Many scientists believe organic molecules owe their existence to interstellar dust. Many molecules were, in fact, first discovered in these clouds. The Moscow physical chemist V. I. Goldanskii wrote in October 1977: 'It seems necessary to consider interstellar grains as possible cold seeds of life, with chemical reactions in the bulk of these grains as a very important stage of prebiotic evolution'.[10]

The dust grains have an average size of between 0.1 and 1 micrometre (a micrometre is one-thousandth—10^{-3}—of a millimetre), their mass is a ten-thousand millionth (10^{-10}) of a milligram, and their density in the interstellar dust approaches what in a terrestrial environment would be an excellent vacuum. In a cube with an edge of 100 metres we would come upon only 20 of these fine dust particles. Only in the dark clouds often silhouetted against bright hydrogen regions, as in the case of the Horsehead Nebula (see Figure 3), does the density of the dust rise by as much as a hundredfold. The tiny grains have a core of graphite, silicon compounds and iron, embedded in a mantle of frozen gases—principally methane, water and ammonia—condensed on to the dust in the low temperatures or formed directly on the dust. According to the 'cold seeds of life' theory, the more complex molecules are formed on the mantle through the stimulation of the light of nearby stars and the catalytic effect of the surfaces of the grains. These reactions will continue at interstellar temperatures—the dust grains in a dark cloud have a temperature around 5 to 15 °C above absolute zero.

Here the newly formed molecules encounter a situation which could quickly wipe them out again. The stars of the Galaxy bathe interstellar space with radiation between longwave radio and far ultraviolet light. (To be precise, down to a wavelength of 91.2 nanometres: at shorter wavelengths the radiation ionizes the hydrogen in its vicinity, and this process absorbs the radiation.) Besides this, the high-energy particles of cosmic radiation penetrate to the depths of the dark clouds. All these effects conspire to destroy the new molecules quite quickly—usually within a century. (Only the molecule of carbon monoxide might survive a little longer, up to 1000 years.) Here only the dust itself can offer protection: if the dust intercepts the destructive ultraviolet radiation (as happens in a dark cloud), the fragile molecules will be adequately shielded, and their life expectancy rises to as much as 10 to 100 million years.

But we have still to establish whether the molecules observed in the dark clouds originated there or possibly elsewhere. It had been suggested that the molecules

Figure 3 The Horsehead Nebula is a dark cloud of dust, which absorbs the light from the luminous hydrogen clouds behind it.

were formed in the outer atmospheres of cool stars and then slowly migrated into the dark clouds. However, this suggestion may now be disregarded, for molecule-rich clouds have also been found so far from hot stars that molecules could not have survived the long journey without protection. Thus only the formation of molecules within the clouds themselves comes in question. Through the catalytic action of the dust grains, single hydrogen atoms (H) combine to form hydrogen molecules (H_2), which then liberate themselves from the dust. These are principally observed in the dense interstellar clouds, with up to several thousand particles per cubic centimetre.

Low-density clouds, typically about 10 particles per cubic centimetre, can still be penetrated by radiation. Frozen gases can evaporate from the grains under the influence of the radiation and thus react together chemically—but the radiation destroys again the molecules it has helped to form. Thus, only one-thousandth of the hydrogen in these clouds is found in molecular form. However, these few hydrogen molecules readily co-operate in the building of other molecules. It is a complex process which produces the simplest diatomic molecule methylidyne (CH) and the methylidyne ion (CH^+), poorer by one electron. First a carbon ion (C^+) and a hydrogen molecule collide, emit the collision energy in the form of a photon, and remain united as the intermediate product (CH_2^+). Further reactions with electrons and hydrogen molecules eventually result in CH and CH $^+$.

Building on these reactions and others, the Harvard University chemists John Black, Alexander Dalgarno, and Michael Oppenheimer explained the observed concentrations of both molecules in low-density clouds in 1975. Radiation is also important when oxygen atoms (O) unite with hydrogen on the surface of the grains to form the molecule OH. Still other diatomic molecules are formed in subsequent reactions with CH, CH $^+$ and OH and elements in the dust.

How are the triatomic and polyatomic molecules, which are primarily seen in dense clouds, formed from these? In the dense dark clouds, radiation is effectively screened, and only the high-velocity protons of cosmic radiation penetrate. Protons, colliding with hydrogen molecules with an energy of 100 mega-electron volts, spark off the reactions crucial to the formation of triatomic molecules. Larger molecules still present us with difficulties. When trying to account for the existence of more complex molecules, chemists can easily lose their way among the rapidly growing number of theoretically possible reactions. But we are not without clues: interstellar chemists were struck by the fact that atoms of nitrogen and oxygen never appear in direct association in interstellar molecules. All searches for compounds such as NO, N_2O, HNO and HNO_3 and more complex molecules with an $N-O$ bond have so far proved fruitless. Another pointer is that, despite intensive searches, no ring molecules have been found—probably an indication that the more complex molecules are formed in the gaseous state where rings of atoms rarely appear, and not on the surface of the grains. But these statements regarding this still young field of interstellar chemistry are by no means final, and the researchers anticipate that improved radio telescopes will lead to further discoveries. The provisional conclusion seems to be that molecular concentrations diminish only slowly with increasing complexity. At present there is no telling what complexity interstellar molecules may achieve.

The denser the dust and gas clouds, the more does the chemistry alter, and the more reactions can proceed concurrently. In the following sections we shall follow the chemical destiny of a condensing cloud up to the formation of a star and its planets. The interstellar cloud whose history we know best is, of course, the one to which the Solar System owes its existence.

A supernova and its aftermath: the Solar System

The central idea of how the Sun and its planets might have been born goes back to 1644 and the philosopher and mathematician René Descartes. Later, in 1755, it was developed by Immanuel Kant and 41 years later by Pierre Simon de Laplace, and became known as the Kant–Laplace theory. According to the modern version of the Kant–Laplace theory, a cloud of gas and dust about a light-year in diameter contracted at the outer end of one of the curved spiral arms of the Milky Way about 4600 million years ago. As the cloud contracted under its own gravitation, its rotation increased more and more. (This results from the conservation of angular momentum: a rotating body compensates for a reduction in radius by increasing its speed of rotation, just as an ice skater spins more rapidly when he pulls his outstretched arms down to his sides.) The cloud thus took on the shape of a rotating disc, giving the primitive Solar Nebula, cold at the edges, hot in the centre. At some point a body formed at the centre, within which, once a critical temperature and density had been reached, nuclear reactions started, thus marking the birth of a star, the Sun. According to Laplace a series of rings then formed in the remains of the disc, from which the planets arose by condensation.

There is, however, no general agreement on the details of this history. Until recently it was commonly supposed that the Solar Nebula condensed out of an interstellar dark cloud. But for this at least three difficulties must be overcome: the clouds are too hot, their magnetic fields are too great, and they rotate too fast. Each of these properties would be sufficient in itself to prevent the collapse of the cloud and the formation of the star. Consider, for example, the rotation: because of the rotation of the Galaxy, which falls off towards the edge, the interstellar clouds are rotating slowly on themselves. As a cloud contracts under its own gravitation, it rotates faster, and centrifugal force increases. As soon as this has grown as large as the gravitational force, the cloud will no longer contract. The magnetic field presents a similar case: the material of the star is only slightly magnetized, but the interstellar magnetic field has a strength of about three-millionths (3×10^{-6}) of a gauss (for comparison, the Earth's magnetic field has a strength of about 0.5 gauss). When a gas cloud shrinks, it carries the built-in magnetic field with it, but this intensifies the magnetic field up to the point where the rotation is brought to a stop. 'It is as if the gas became elastic.'[11]

There is, however, one mechanism by which the cloud can dispose of both magnetic and rotational energy at the same time. If the magnetic field of the cloud is coupled with the galactic magnetic field, the gas in the cloud remains linked to its surroundings. The accelerating rotation of the cloud drags the fields with it, and these, like rubber bands, provide increasing resistance. This has the desired effect; it brakes the rotational energy and reduces the field strength of the rotating masses.

It may seem odd that the heat of the cloud also presents a problem. When an interstellar cloud condenses to a stage that can later develop into a star, the number of particles per cubic centimetre rises from a few thousand to millions or thousands of millions. The temperature correspondingly rises to 1000 °C—clouds with several times the mass of the Sun have been observed by their infrared radiation. Rising temperature and increasing density bring about a greater gas pressure from within, outwards, which slows down the contraction of the cloud. If the cloud heats up too much, the gas pressure may completely stop the contraction due to gravity: the cloud would then dissipate once more. The heat energy of a cloud is in fact the greatest difficulty of the three.

The initial development of the cloud, then, is still a matter for dispute. Only the point in time at which the Solar Nebula separated itself from the rest of the interstellar gas seems established: about 100 million years before the formation of the planets, 4700 million years ago all told. This can be determined from the abundance of long-lived radioactive material. The concentrations of uranium-238 (half-life 4500 million years) and of thorium-232 (half-life 20 000 million years) have been measured in meteorites and in rocks from the Earth and the Moon, and indicate that the planets developed a solid crust 4600 million years ago.

Next, the radioactivity prevailing on the planets at the time of their formation is compared with the radioactivity of the interstellar gas. The interstellar gas too is radioactive, for it has been repeatedly enriched with radioactive elements by stellar explosions. From this comparison, using in particular the elements plutonium-244 (half-life 80 million years) and iodine-129 (half-life 16 million years), the time interval mentioned above—100 million years—can be calculated. The elements themselves, it is true, have long since disintegrated, but they can be traced through their decay products.

Controversy also surrounds the mechanism by which the primitive Solar Nebula became sufficiently compressed to undergo a final contraction—despite heat, magnetic field and rotation. In 1977 there was new evidence that a nearby supernova explosion had brought about the contraction of the primitive Solar Nebula.[12-16] As early as 1953, E. J. Öpik had put forward the hypothesis that compression by shock waves could initiate the formation of stars in gas clouds, but until recently this theory lacked verification. Physicists have now confirmed the theory, following thorough analysis of the composition of several meteorites, for instance the Allende Meteorite which fell in Mexico on 8 February 1969. Meteoroids, wanderers among the planets, can be regarded as the visiting cards of the Solar System announcing its birth; having arisen in the primitive Solar Nebula, they reflect its chemical composition. Investigating teams from the California Institute of Technology (CalTech) and the University of Chicago observed that certain isotopes were present in higher proportions in meteorites than on Earth. In particular, oxygen-16, magnesium-26 and neon-22 exceeded their terrestrial abundance by up to 5 per cent. The magnesium-26 which was found is regarded as a disintegration product of radioactive aluminium-26, which has the 'short' half-life of only 120 000 years. But how did the aluminium-26, 4600 million years ago, get into the newly formed meteorites? Because of its short lifetime, it must have been propelled into the primitive Solar Nebula while still 'fresh', probably by the shock wave of a nearby supernova. The pressure of the shock wave then compressed the molecules and dust particles enough for their mutual gravitation to hold them together. From then on the cloud contracted slowly. The primitive nebula of the Solar System had become detached from the masses of galactic gas. The short lifetime of the aluminium isotope indicates that the supernova outburst can have taken place only a few million years before the initial formation of the Solar Nebula.

While these results were still being discussed, further support came from the Eighth Moon Conference, held in March 1977 in Houston (Texas). The astronomers William Herbst and George E. Assousa (Carnegie Institute) appeared to have found a similar case in the constellation Canis Major. In the young star cluster CMa R-1 they came upon a gaseous shell, a supernova remnant, with a diameter of 190 light-years. The astronomers reconstructed the age of the remnant as follows: with the present expansion velocity of 32 kilometres per second, the explosion must have taken place about 700 000 years ago. But the stars in the cluster are of exactly this age. Herbst and Assousa consider that these observations have confirmed the process of supernova-induced star formation and thus strengthened the argument that the Solar System was triggered by a supernova.

The birth of the planets

Many of the properties of the Sun and the planets, even as we see them now, are still not understood. A complete theory of the origin of the Solar System must furnish an explanation for all its components: the single large body, the Sun, and its set of smaller satellites: the four small inner planets, Mercury, Venus, Earth and Mars, as well as the five large outer planets, Jupiter, Saturn, Uranus, Neptune and Pluto. Added to these, there is a swarm of at least 100 000 million (10^{11}) irregular comets, which surround the Solar System at about 100 000 times the distance of the Earth from the Sun, forming what is known as the Oort cloud. Three of the larger planets, Jupiter, Saturn, and Uranus, themselves present miniature solar systems, with groups of inner and outer moons plus irregular satellites. Nevertheless, the fraction of the total mass contributed by the planets and comets is vanishingly small: taken together, they provide only 1.4×10^{-3} of the mass of the Sun, and of this the largest part is represented by Jupiter and Saturn. The combined mass of the comets is estimated at several hundred Earth masses.

A plausible but still controversial scenario for the origin of the Solar System is offered by the theory of the Harvard scientist A. G. W. Cameron.[11] The primitive Solar Nebula contained about 1 solar mass and was as thick as the Earth–Sun distance. By some tricky process, very early in its life, the Sun must have lost almost all its rotational energy, for although it provides 99.99 per cent of the mass of the Solar System, it possesses only 2 per cent of its total angular momentum. The greater part of the angular momentum of the gas that went to produce the Sun must therefore have been transported away. Of course this is only possible if a considerable part of the primitive Solar Nebula remained at a great distance from the Sun, so as to take up this angular momentum. The denser and hence more rapidly contracting inner parts evolved a smaller primitive Solar Nebula, towards which matter continued to flow inwards from the remaining cloudy envelope. By this inflow of matter the angular momentum from the centre of the primitive Solar Nebula—the future Sun—was transported outwards against the flow of matter and carried to the cloudy envelope. The primitive Solar Nebula took on the form of a massive centre poor in angular momentum, surrounded by a disc of gas and dust.

The planets began to form when the dust particles (diameter 1 micrometre), accelerated by gas turbulences, collided and formed grains 1 centimetre in size. As Peter Goldreich (CalTech, California) and William R. Ward (Harvard) have recently suggested, asteroids as much as 1 kilometre across could result from these once the grains had migrated to the central plane of the disc. By the same mechanism, these asteroids became grouped in loose clusters, which could collide with one another in the same way.

In their further development the inner and outer planets go different ways. Only the inner planets could have arisen from the amalgamation of such clusters. For Jupiter and Saturn, which consist principally of hydrogen and helium, gas from the primitive Solar Nebula must have been concentrated about planetoid nuclei until such time as it collapsed on the cores of the planets. On this theory the asteroid belt, interposed between the small inner and massive outer planets, between Mars and Jupiter, is not a destroyed planet, but one which was never born, its formation prevented by the proximity of Jupiter (with 318 times the mass of the Earth).

Once the gas density at the centre of the primitive Solar Nebula had exceeded a certain value, hydrogen nuclei began to combine and the Sun began to shine. But shortly after their birth young suns pass through a stage, called the T-Tauri phase, in which they blow off large amounts of matter. This material could sweep through the Solar System like a powerful solar wind, and thus clean it of the residual gas. The comets, dirty snowballs of ice and dust, were probably formed within the Oort cloud: smaller cloud fragments circling the primitive Solar Nebula contracted (like the nebula itself) to a disc, and analogously to the planets produced the numerous small comets.

But many phenomena remain unexplained and we are still far from understanding all the properties of the planets. However, there is more certainty about the origin of planetary material—the planets were formed from the unaltered substance of interstellar matter. This idea, too, was for a long time a source of controversy among scientists; as late as the 1960s a hypothetical alternative lay in the possibility that the planetary material, like the solar gas, had been heated to stellar temperatures, and had thus undergone significant changes in its chemistry, before it cooled off and condensed to form the planets. This was definitively refuted by astronomical observations in 1975 and 1976. At that time the abundance of deuterium (heavy hydrogen) relative to hydrogen was measured in the interstellar gas; it was found that two deuterium atoms were encountered for every 100 000 hydrogen atoms. The same value was obtained in

Figure 4 In 1977, using an infrared telescope carried on a plane, American astronomers discovered a new source in the constellation Cygnus, in which a planetary system may just be coming into being. The data may be interpreted as indicating a hot disc of dust, at whose centre a young star is just being born. The radiation may be produced by gas and dust falling out of the disc into the star. A number of planets may be formed in the disc during this phase. In the illustration, the cooler outer regions are represented by darker shades of grey. (Courtesy: NASA.)

Jupiter's atmosphere; analogous data were also obtained for the element lithium.

For Solar System astrophysicists, 1977 offered something of a bonanza. Studies of meteorites verified the hypothesis that a supernova compressed the primitive Solar Nebula. At the same time it was shown that exactly the same process in the constellation Canis Major assisted at the birth of other stars 700 000 years ago.

Planetary theory had its successes, too. A group of astronomers from the Steward Observatory of the University of Arizona and from the NASA Ames Research Center discovered a star within a disc surrounded by a cloud of hydrogen gas and dust. From the results, published in November 1977, it can be concluded that matter from the disc is flowing inwards to the central star, exactly as planetary formation theorists expect.[17] The object in question is MWC 349 in Cygnus (see Figure 4).

The new star is about 10 times as large and 30 times as heavy as the Sun. The disc has an observable diameter 10 to 20 times that of the central star. Near this central object, as one of the discoverers explains, the disc is hot and rotates faster; further out it is cooler and thus harder to observe. In the cooler part of the disc the conditions for the formation of molecules and fast-growing dust grains are ideal. 'Although this does not necessarily imply formation of larger masses such as planets, the physical circumstances seem favourable to such a process', say the astronomers. Constrained by friction, the greater part of the hydrogen and dust will flow towards the centre of MWC 349, and will continue to do so until nuclear burning processes are initiated in the central body as it grows denser and hotter. Other material drifts outwards from the system under the pressure of radiation. But larger condensations, large enough to withstand the forces of friction and radiation, may continue orbiting the central star of MWC 349 as planetary nuclei. There are as yet no further observations to strengthen this suggested interpretation. For the present, we must also consider other ways in which the observed radiation could possibly have been produced—perhaps in a *spherical* dust shell rather than a disc, with a flow towards the central star. Thus, this problem is still in need of detailed analysis.

Within quite a short time, in a few million years, MWC 349 may be a young star, surrounded by a whole troop of large and small planets. Will the arrangement of these planets be similar to that in the Solar System? In 1970, the American planetary physicist Stephen H. Dole undertook a computer simulation of the development of the planets from the primitive Solar Nebula, varying the initial distribution of matter in the nebula by a randomizing process.[18] For each (numerically expressed) initial distribution in the primitive nebula the program then calculated the final distribution of the planets. The results showed that there is every possibility that a system similar to our own may develop. Figure 5 shows the Solar System as it is: the masses of the planets are represented by the sizes of the circles; the double bracket beneath each planet denotes the smallest, the mean and the greatest distance of the planet from the Sun. Below the diagram of the Solar System are some of the 'best' computed alternatives—those most resembling the Solar System. It is evident at a glance that the arrangements of the planets are not too different from that in the Solar System. It is striking that four or five low-mass planets always form in the neighbourhood of the Sun, followed by one or two Jupiter-sized giant planets. This is also true for other variants, not shown here, in which the arrangement of the planets diverges further from the actual distribution. In these systems some planets are set closer together, with the possible danger of later collisions. But despite the relative similarity of the simulated alternatives, in only one case does a suitable planet fall within the range of distances that can be regarded as habitable according to the most recent researches (see 'The Earth's atmosphere: minimum conditions for life', page 27). Dole's studies impressively confirm the condensation theory of the planets. But what Dole's program does not cover is the origin of the Moon and the

The planets of our Solar System

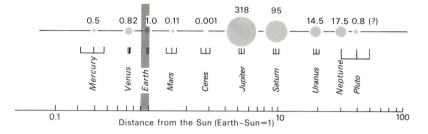

Simulated solar systems

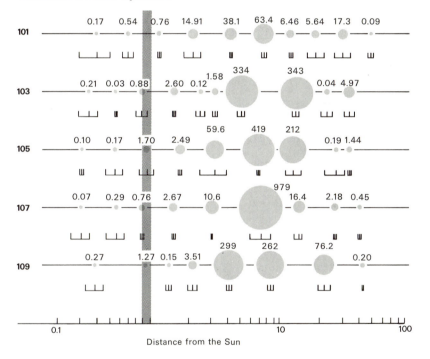

Figure 5 The Sun's planetary system and some computer-simulated variants. The sizes of the circles indicate the respective planetary masses; the accompanying figures give their masses in terms of the Earth's mass. The double brackets under the planets show their smallest, mean and largest distance from the Sun. The numbers at the left are random values; these determine the points at which condensation takes place with the Solar Nebula. The habitable zone or 'ecosphere' of the Sun is shown shaded at about the Earth's distance. (See ref. 18.)

satellites of other planets. He considers, however, that this can also be elucidated within the framework of the theory he employs if more detailed processes are incorporated into the program.

When the Sun dies . . .

When the Sun swept from the Solar System its light gases and the dust left behind by the planets, it also came to the end of the embryonic stage of its development. While previously it had made an appearance only as a hot cloud of dust, in whose interior the first nuclear reactions were taking place, it now presented itself as a young star, surrounded by a system of planets and comets. The fusion of hydrogen to helium (see Figure 6) marks the longest and most stable period in the life of a star of solar mass, comprising 99 per cent of its total life span. During this time the enormous energy output from the fusion process heats up the gas, so that the resulting gas pressure and the radiation pressure together balance out the force of self-gravity, and thus prevent further contraction. Today the Sun is about 5000 million (5×10^9) years old, and has reached about the halfway point in its life expectancy.

But like all stars, the Sun will one day die. However, the lifetime of a star depends on its mass, and as a rule of thumb we may say: the heavier a star, the shorter is its life (see Table 2). Heavy stars consume their nuclear material faster than less heavy stars. Even at 10 times the solar mass, nuclear burning proceeds 1000 times faster, thus shortening the lifetime of the star to 100 million years. On the other hand, small stars have a future that can exceed the present age of the Universe: a star with a tenth of the solar mass can shine for a million million (10^{12}) years, though with a feeble light.

When the hydrogen in the star's interior is used up, the beginning of the end is near. As a result the site of hydrogen burning moves towards the outer regions of the star. This forces the outer layers of the star further outwards until a new equilibrium is established. The star swells to 100 times its size and takes on a reddish colour: it becomes a 'red giant'. When the Sun reaches this stage, it will engulf Mercury and Venus and even the Earth. The temperature of the outer layers of the red giant Sun will be about $3000\,^\circ$C; the Earth's atmosphere will be torn away and the Earth's surface burned up. There are two possibilities for the further destiny of the Earth. Either the charred planet will slowly spiral into the interior of the swollen star, or it will be driven outwards by the expanding solar gases.

After about 1000 million years the Sun ends its existence as a red giant (see Figure 7). The remaining hydrogen in the stellar envelope is now used up and the most fertile source of stellar energy is exhausted. The consequences are drastic: without an adequate supply of energy the temperature falls, and with it the gas and radiation pressure. Gravitation gets the upper hand, and the star shrinks

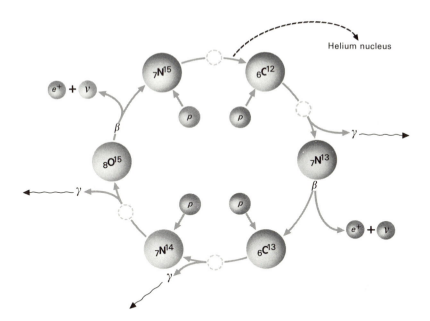

Helium nucleus

Figure 6 Fusion of the nuclei of hydrogen atoms (protons) to helium nuclei (alpha particles). The carbon–nitrogen cycle shown here schematically is a closed sequence of nuclear processes suggested by Bethe and von Weizsäcker. This is the process by which the Sun produces energy in its interior. The symbols represent: ◯, transition product; p, proton; β, beta decay; γ, gamma decay; e^+, positron (anti-particle of the electron); ν, neutrino. With the assistance of carbon (C), nitrogen (N) and oxygen (O) atoms, four protons are consumed on each cycle to produce one helium nucleus. (After ref. 2.)

Table 2 Vital statistics of the commonest classes of stars

Class	Mass (Sun = 1)	Temperature of star (°C)	Lifetime of star (10^9 years)	Abundance (%)
O	32	35 000	0.001	0.08
B	6	14 000	0.008	0.5
A	2	8 100	0.81	0.81
F	1.25	6 500	4.2	4.2
G	0.9	5 400	10	10
K	0.6	4 000	32	32
M	0.22	2 600	210	47

(The classifications of the stars is drummed into astronomy students with the following mnemonic: *Oh Be A Fine Girl, Kiss Me (Right Now)*—this sequence was at one time mistakenly thought to represent the progress of stellar development.)

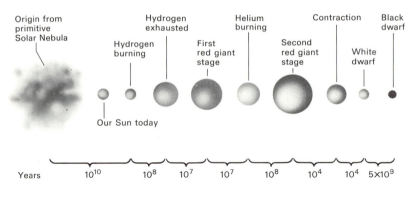

Figure 7 Life history of the Sun.

until the temperature in the centre rises to 200 million °C. This is high enough to kindle the second stage of fusion: the fusion of helium nuclei into carbon. This gives the star its second wind: it promptly expands again and appears once more as a red giant. The kindling of helium fusion takes place so fast (within seconds) that astronomers have called this transition the 'helium flash'.

At this point the star's fate reaches a parting of the ways. If the star was less than about 1.5 times the mass of the Sun, the second red-giant stage ushers in the final step. If it contracts now, it does not liberate enough gravitational energy to initiate the third stage of fusion—the conversion of carbon into heavier elements. The star grows unstable, and its outer layers are flung into space. The heavy core is left behind, a sphere now only a few thousand kilometres in diameter, a so-called 'white dwarf'. It is extremely dense—1 cubic centimetre 'weighs' several tonnes (density 10^6 grams per cubic centimetre), and its surface gravity is correspondingly high. What keeps a white dwarf in equilibrium is no longer gas pressure fed by the energy of fusion, but a quantum-mechanical effect: the *degeneracy* of the electron gas within the star. By this effect the electron gas exerts the pressure required to counterbalance gravity. Hotter than the Sun, but only the size of the Earth, the white dwarf can doze away its few remaining thousands of millions of years of stellar senility. It slowly cools and ends as a barely observable 'black dwarf'.

Heavier stars, of more than 5 solar masses, suffer a dramatic fate: through several cycles of expansion and contraction the heavier elements magnesium, silicon, cobalt, nickel and so on through to iron are formed in their interiors. For some of these stars the red-giant stage has a catastrophic end: in a mighty explosion, a 'supernova', the outer parts of the star are blown away. For a short time the brightness of a supernova may even exceed that of the whole galaxy in which it occurs. Supernovae briefly heat up to such a temperature—thousands of millions of degrees C—that even the heaviest elements, as far as uranium, are produced. When the centre of a supernova contracts under the shock of the implosion, a condensed central body can be left, which consists principally of the nuclear constituents neutrons and protons, appearing as iron in its envelope—a star about 20 kilometres in diameter, a *neutron star*. It is more than 10 000 times denser than a white dwarf. Now it is the degeneracy of the neutrons which maintains equilibrium with gravity. Neutron stars are as dense as atomic nuclei, and may rotate many hundred times a second. The electrons from the stellar surface are accelerated along the axis of its dipole magnetic field, and consequently emit radio waves. We receive these waves on the Earth as they briefly shine on us once during each rotation of the star, in rather the same way as the beam from a lighthouse is seen intermittently. These stars are the *pulsars*, first discovered in 1967. On the surface of a neutron star we find the strongest magnetic fields known in the Universe: a research group from the Max Planck Institute for Extraterrestrial Physics in Garching first established an exact value of 7000 gauss by a balloon flight in 1977.[19] The remains of the supernova of 1054, the Crab pulsar, is still rotating today at the centre of the Crab Nebula. Each second it turns 33 times about its axis.

Let us now turn away from stars to the environments they must create on planets for life to evolve. The history of the Earth's surface naturally gives us the best-known example of this.

The Earth's atmosphere—
the minimum conditions for life

Initially the newly formed Earth was very hot, the temperature at the surface rising to about 1500°C. While the Earth was still condensing into a planet, its innermost regions melted, and the heavier metals such as iron and chromium sank to its centre, while the light elements, especially silicon, rose to the surface like fat on soup. But the heavy radioactive elements behaved otherwise. The electrons in their atomic shells move in higher, 'excited' orbits, further from the atomic nucleus. Hence the effective volume of radioactive atoms is increased, and they take up relatively more room. Although uranium is heavier than iron, it thus rose to the surface along with other radioactive elements and settled in the Earth's crust. In the end, little radioactive material remained at the centre; most of it still occurs today in the lower strata of the Earth's crust.

At this stage in the Earth's history, a primitive and probably fairly dense atmosphere, consisting of carbon monoxide, hydrogen and nitrogen, formed from condensation. Other molecules, such as methane and ammonia, would have been destroyed by the temperature of 1500°C. Later the Sun awoke and went through its T-Tauri stage, during which it ejected large masses of gas. This gas swept through the Solar System at great speed and, as it were, 'shaved' the inner planets of everything they had accumulated in the way of a primitive atmosphere during the condensation process. But this did not leave the Earth breathless—volcanoes poured fresh gas from the Earth's interior, and formed a new, second atmosphere about the planet.

The work of Michael H. Hart, at NASA's Goddard Space Flight Center, will describe for us the further development of the atmosphere after this point. In 1977 he made a study of the development of our present atmosphere from the (secondary) primitive atmosphere, using computer models. His startling conclusion is that if the distance of the Earth from the Sun at the time of the formation of the Solar System had been only a little greater or a little less, the development of life on Earth would to all intents and purposes have been suppressed for all time.[20-22]

Hart's first problem was to establish the composition of the secondary atmosphere. In this he was partly helped by the radioactive elements in the Earth's crust. Their half-lives are known, which means that we can calculate back, on the basis of the abundance of the still active radioactive material found today, to the abundance that had accumulated in the Earth's crust some thousands of millions of years ago. When we have thus determined the composition of the Earth's crust at the time when the Sun had just completed its T-Tauri phase, it is possible to calculate the abundance of gaseous fission products liberated by radioactive disintegration.

Calculation of these processes on the computer shows that carbon dioxide must have predominated in the early atmosphere, along with water vapour and a few per cent of noble gases such as argon, produced by the radioactive disintegration of potassium-40. Geologists consider that the amount of gas ejected into the atmosphere died away exponentially over the following 800 million years of the development of the Earth's climate. The most important external factor governing changes in the atmosphere is now the Sun. In the last 4000 million years its surface temperature has risen by about 300 °C to the present value of about 5500 °C, as a result of its development into a 'normal' star in the main sequence, the largest group, comprising stars rich in hydrogen. Thus, at the time when volcanoes were still forcefully replenishing the atmosphere, the Sun was radiating about 25 per cent less energy towards the Earth.

This circumstance has always presented climatologists with a problem. Because the Sun was cooler at that time, the temperature on Earth should have been so low that all the oceans would have turned to ice. Some mechanism must have kept the Earth warm enough until the Sun itself got properly under way.

Hart's calculations show how Nature may have solved this problem. At this time the Earth's surface was somewhat warmer than could have been achieved by the Sun alone: it was only possible with the help of the Earth's atmosphere. The key lies in the 'greenhouse effect', a consequence of the carbon dioxide content of the atmosphere. The molecules of carbon dioxide, and to a lesser extent those of ammonia, operate as a filter within a definite range of wavelengths. They possess the property of allowing large amounts of sunlight to reach the Earth's surface, but of preventing the re-radiation of reflected energy, transformed into infrared (heat) rays. An atmosphere of carbon dioxide operates not unlike the glass roof of a greenhouse. The greenhouse effect caused the mean annual temperature to rise to over 40 °C: on the average it was about twice as warm as today. Large quantities of water evaporated from the oceans, so that clouds completely covered the Earth. Without the greenhouse effect in the early days of the Earth's development, there could have been no life on our planet. The seas, frozen to ice, would have reflected almost all the incoming solar radiation back into space; thus the energy which might eventually have melted the ice would not have been accumulated, and even today, despite the increasing solar radiation, it would not have melted. The large carbon dioxide content of the secondary primitive atmosphere thus helped to remove a crucial barrier to the origin of life. The greenhouse effect warmed the Earth for about 2500 million years (see Figure 8).

Gradually the carbon dioxide content of the atmosphere was reduced, mainly by silicate compounds, in the so-called Urey reaction, named after the planetary chemist Harold C. Urey. There remained ammonia and methane, which for a time maintained the greenhouse effect. In the upper layers of the atmosphere, sunlight destroyed the remaining water vapour molecules, splitting them into their component parts, hydrogen and oxygen (photolysis). Part of the hydrogen escaped into space. Some oxygen was also contributed, 3500 million years ago, by photosynthesis carried out by the first scanty plant life. But the oxygen immediately combined with the iron contained in the surface minerals and turned it to rust. At the same time it also reacted with methane and ammonia, converting them to water and nitrogen. So for the time being the overall oxygen content remained low. Then, about 2000 million years ago, the further evolution of life almost came to an end. After the carbon dioxide had been almost entirely removed by the Urey reaction—the carbon dioxide abundance fell below the 5 per cent mark—the greenhouse effect, previously such a significant factor, diminished. The clouds which had hitherto covered the whole Earth fell as rain and the weather grew clear and dry. Thus the Earth lost the heat buffer on which it had depended. The Sun, although warmer than 2000 million years previously, had still not reached its present-day radiation intensity, and the temperature inevitably fell steeply. It would appear that at this point the Earth and the evolution of life escaped by the skin of their teeth.

The temperature fell to the lowest mean value in the entire history of the Earth: at the surface the yearly average was under 10 °C. Large ice caps grew at the poles. Only a few degrees less, and all the seas would have frozen completely into an icy desert. The fall of temperature and the radical change of climate presented a dangerous hurdle on the road to life. A very slight difference in the Earth's orbit could have caused the temperature to fall below a critical value. If the Earth's orbit at that time had lain on the average about 1 per cent further from the Sun, the mean temperature would have fallen to freezing point. The oceans would have frozen, a full-blown ice age would have set in, and further solar energy

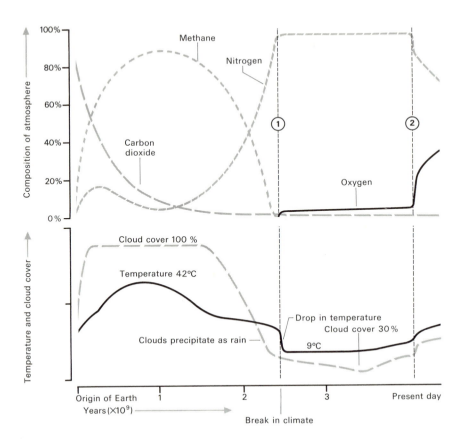

Figure 8 Evolution of the Earth's atmosphere up to the present day. The mean surface temperature of the Earth fell sharply 2000 million (2×10^9) years ago, according to the calculations of Michael H. Hart (line 1). At the same time the clouds precipitated as rain. Previously the greenhouse effect had driven the temperature up to more than 40 °C, thus evaporating large amounts of water from the oceans to cover the Earth completely with clouds. From this point, also, the oxygen content slowly rose. Once there was enough ozone to shield the ultraviolet radiation of the Sun so as to make the Earth's surface hospitable to plants, the highly efficient process of photosynthesis caused the oxygen content to rise sharply (line 2). (Curves from ref. 20.)

would have been reflected away by the bright ice. The situation would then have taken on a permanent resemblance to the weather on Mars.

In the other direction too, towards the Sun, the tolerance was fairly small. A 5 per cent smaller distance from the Earth to the Sun would have condemned the Earth to perpetual sleep in an overheated greenhouse atmosphere. On this point Hart's calculations confirm an earlier study of the scientists S. I. Rasool and C. de Bergh, who had already arrived at this result in 1970. Let us return for a moment to the secondary primitive atmosphere, fed by gases issuing from the

Earth's interior. Carbon dioxide dominates, clouds of water vapour enshroud the Earth, and some methane and ammonia are present. Only if the radiation from the Sun does not exceed a certain limit, that is, if the planet does not orbit too close to the Sun, will it remain cool enough for water vapour to condense; the carbon dioxide will then be destroyed via the Urey reaction and dissolve in water. If the Earth follows a path closer to the Sun, however, the greenhouse effect will remain so strong that barely any water will condense out and the full cloud cover will persist. But with less water in the seas less carbon dioxide is destroyed (the Urey reaction proceeds best in water), and the carbon dioxide content will increase.

This further strengthens the greenhouse effect: the surface temperature rises, more water evaporates, the Urey reaction is even more hampered. As the Sun's radiation has increased steadily during the last 4000 million years, this process could never have been reversed.

The Earth orbits the Sun at an average distance of 150 million kilometres. If its orbit were only 6 million kilometres smaller, the Earth would have shared the fate of Venus—a carbon dioxide hell, with sulphuric acid gales and a surface temperature of 500 °C. 'The evolutionary process is very sensitive to the Earth–Sun distance,' says Hart. Since at the time of the climatic change almost all the iron was already oxidized and all methane and ammonia had reacted with oxygen, the oxygen content increased steadily from then on. Until the atmosphere had been sufficiently enriched with oxygen and ozone, the deadly ultraviolet rays of the Sun rendered the continents inimical to life. And as adequate protection from ultraviolet radiation is only guaranteed by a depth of at least 10 metres of water, an important zone of the oceans was unsuited to the rise of life.

The ozone layer that shelters us today lies at a height of about 30 kilometres. But it also stretches 'fingers' down to the flight altitude of passenger planes, occasionally producing symptoms of poisoning in the passengers and headaches for the airlines. It must be filtered out in the ventilation or destroyed by heating. The ozone content of the atmosphere is more or less constant. It is only slowly replenished by Nature; but it is very rapidly destroyed by ascending compounds of fluorine and chlorine, which are used as propellants in aerosol sprays. With less atmospheric ozone, more people will suffer from skin cancer, and for this reason some states in the USA are already beginning to ban these aerosols, as has recently happened in Sweden.

But at that time, fortunately, aerosols had yet to be introduced, and life could develop undisturbed in the shelter of the growing ozone filter. Not until 420 million years ago did the oxygen and therefore the ozone content become high enough to screen off the ultraviolet radiation to such an extent that the Earth's surface became endurable for living things.

But we must not get ahead of ourselves. Let us first examine the reliability of Hart's results and consider their implications for the chances of the evolution of life. On the one hand they agree with the previous limited findings; on the other hand Hart was able in his computer program to vary almost all the processes that influenced the sequence of events—the quantity of volcanic gas issuing, the condensation of water vapour in the oceans, the dissociation of water vapour in the upper atmosphere, the escape of hydrogen into space, chemical reactions between the various gases, the influence of primitive organisms, the extraction of carbon dioxide from the atmosphere by minerals and the slow increase in the Sun's strength. Many times Hart repeated his simulations, always with different percentage abundances of carbon dioxide, oxygen and nitrogen in the gases streaming from the Earth's interior. The program computed the changes in all the components of the atmosphere and the oceans in steps of 2.5 million years. After each step, along with other data, the amount of water in the oceans and the amount and composition of the atmosphere were determined. Various initial mixtures proposed for the second primitive atmosphere led to purely computational atmospheres for the present time which differed from the actual one. Figure 8 shows the development of the atmosphere with time up to the present, starting from the mixture of gases that gave the result nearest to the present atmosphere.

The life-supporting zones of the stars

Although refinements of Hart's programs can still be expected, I consider their most important statement to be correct: *life on Earth clutched at a straw*! The zones around stars in which life can develop—the so-called *ecospheres* of the stars—are considerably smaller than has been thought for some decades. Stephen H. Dole discusses this subject in detail in his book, *Habitable Planets for Man*, a second edition of which appeared in 1970.[23] For Dole, the ecosphere is 'a region in space, in the vicinity of a star, in which suitable planets can have surface conditions compatible with the origin, [and] evolution of life to complex forms'. Dole further demands 'the existence of land life *and* surface conditions suitable for human beings'. Before Dole the NASA scientist Su-Shu Huang had concluded that the ecosphere of the Sun extends—if we set the Earth–Sun distance as 1.0—from 0.7 to 1.3.[24] On this basis life similar to that on Earth would have been possible on Venus (0.72) and almost even on Mars (1.52). Dole limits the ecosphere of the Sun to distances between 0.785 and 1.24. But if Hart's calculations are correct, we must now fix the limits of the ecosphere to the range between 0.95 and 1.01! Compared with Cameron, this represents a reduction by a factor of 10, and set against Dole a factor of almost 9.

And even despite this the Earth struck lucky. For an important role is played by a further circumstance which Hart was first to take into account: the luminosity of the Sun had increased by 25 per cent in the last 4 million years. It appears that we owe our existence not least to this spontaneous heating up of the Sun by 300 °C. In response to my enquiry, Hart replied (1977): 'The question you raised—whether an increase in solar luminosity over geological time is in fact *essential* if a habitable planet is to result—is extremely interesting. Actually, I have been occupied with intensive research into this question for the last few months. My results suggest that the answer is yes, and that a constant (or too slowly increasing) solar luminosity would not have permitted the development of an atmosphere such as we have today. For this reason the continuously habitable zones of stars lighter than the Sun are not only nearer to them, but also narrower, than the continuously habitable zone about the Sun. It therefore seems that [lighter] stars beyond class K0 [where the outer and inner limits of the habitable zone overlap] no longer possess continuously habitable zones.'[25]

The stellar data in Table 2 show that K stars, of lower mass than the Sun, occupy the neighbouring class. (The classes are usually further divided from 0 to 9—for instance, K0, K1, K2 . . . to K9—explaining the 'K0' class in Hart's letter.) K stars are also cooler than the Sun. Before Dole's work, habitable zones were ascribed to all K stars as well as to the very cool M stars, on the basis of temperature. Dole drew the limit of uninhabitability even at the K2 stars. And what of the ecospheres of the massive stars? The giant, high-mass stars of classes O, B, A and F are much hotter than the Sun; their ecospheres are thus further from the star. But these stars have only a short lifetime, and account for only 3.5 per cent of all stars.

No more than 1 per cent of stars fall into the O class with more than 30 solar masses; but because of their high luminosity they appear very striking in the sky. The formation of a habitable planet requires at least 3 million years—assuming that certain other conditions, which we shall shortly discuss, are fulfilled. Given their short lifetimes, stars of types O, B and A therefore fail as candidates for extraterrestrial life. And if we set as a minimal requirement the time that life has taken to evolve on Earth—4500 million years—F stars are closed to civilization. Thus, only the G stars remain, including the Sun (to be precise, it is a G2 star); when we speak in the following of *stars like the Sun*, we will always mean stars of class G. The ecosphere of a G star is largest for the subclass G0, and grows increasingly narrower up to G9, the last subclass.

In order for life to develop, a planet must be formed exactly in this narrow ecosphere. But this alone is not enough. The planet must possess a number of additional properties in order to qualify as a potential bearer of life: its rotation on its axis, its mass, the departure of its orbit from a true circle, the inclination of its axis—even its moons must move within relatively narrow limits of tolerance.

A planet and its ice ages

The mass of a planet is of first importance, for it defines the surface gravity, how the atmosphere is composed, how great the atmospheric pressure is and how high the greatest mountain can be. If the mass of the planet is too small, it can retain no seas and no breathable atmosphere; oxygen would evaporate into space. If, on the other hand, it is too massive, the atmosphere would be too dense, with the possible consequence that only a little sunlight would penetrate to the surface and photosynthesis would be greatly hampered. If we take all these limitations into account, a planet with life like that on Earth must fall between 0.4 and 2.35 Earth masses. This corresponds to a force of gravity on the surface between 0.68 and 1.5g (g represents the Earth's surface gravity).

How fast may the planet rotate? If it turns too slowly it heats up too much on the day side and would be too cold on the night side. An extreme example is the planet Mercury: it turns on its axis only once every 2 months. By day it heats up to 350 °C, while the night side cools off to the unearthly chill of − 170 °C. The faster the rotation, the less the temperature varies but the stormier grows the atmosphere. The upper limit is reached when the equatorial speed causes the planet to shed material into space, or the planet's crust becomes noticeably distorted. In short, by these criteria a planet must rotate on its axis at least once in 4 terrestrial days, and its day must be no less than 3 hours in length.

However, when a planet always presents the same side to its star, as the Moon does to the Earth, the planetary day is of exactly the same length as the planetary year. We must therefore ask whether in this case a zone of life might be possible at the narrow, fixed frontier between day and night. The planetologists say that this situation never arises at the time when the body is formed, but gradually comes about by means of a gradual slowing of the rotation. During this period, all the water would probably be broken down by the effects of radiation, and the resulting hydrogen would be lost. Such a planet is thus uninhabitable: its night side would be frozen, its day side a desert.

The inclination of the planet's axis also has a climatic effect. The more the rotation axis is inclined to the plane of the orbit, the smaller is the zone in which the mean annual temperature varies within set limits. The Earth's axis is inclined by 23.5 degrees. True, 10 per cent of the planetary surface may still be habitable with an axial inclination up to 80 degrees allowable— if at the same time the radiation falling on the planet from the star rises to 1.9 times that falling on the Earth. Thus, in this extreme case we would have to move the planet closer to the star.

Finally we must consider possible changes in the geometry of the orbit. If the path of the planet deviates only slightly from a circle, this has a negligible influence on the mean yearly temperature. But beyond this (with a numerical eccentricity greater than 0.2) the changes of temperature start to have a drastic effect: just how sensitive the Earth's climate is to this characteristic is shown by the ice ages. As James D. Hays of Columbia University has explained, geologists are now certain that changes in the geometry of the Earth's orbit were the cause of the ice ages.[26] For the axial inclination and orbit of the Earth undergo periodic changes. The various periods overlap and lead to long-term changes in the climate.

The *elliptical orbit* of the Earth revolves about the Sun in a certain period. In addition, the rotation axis of the Earth precesses in a 26 000-year cycle. Today the

Earth and Sun are closest in January: in 10 000 years' time they will be closest in July.

The *plane* of the Earth's orbit tilts back and forth with a rhythm of 41 000 years, so that the Earth's axis is sometimes less inclined to the Sun than at other times. It was inclined most steeply 9000 years ago, so that today the axial inclination is decreasing. As a consequence the polar ice caps will grow larger.

Finally, 93 000 years is the period with which the Earth's orbit changes from being practically circular to a markedly elliptical shape and back again. At present the Earth's orbit is about midway between the two extremes, and is slowly gaining greater ellipticity.

As long ago as the 1930s the Serbian geophysicist Milutin Milankovitch had developed a detailed theory of the relationship between ice ages and orbital changes; but at the time there were no accurate long-term data for the Earth's climate. Hays and his team have now provided these data by means of samples of the sea bottom in the Indian Ocean. This permitted the reconstruction of the course of the climate over the last 450 000 years—an interval three times as long as had previously been known. From the abundance of plankton remains in each layer the geologists deduced the temperature at different periods (plankton is very sensitive to changes of temperature). The results verify Milankovitch's theory, and indicate, from the interplay of the variations above, a further ice age, which will be upon us in some 10 000 years' time.

Although Hays's analyses provide a plausible explanation for climatic events of the past few million years, they do not explain why there were no ice ages before that. If there is a connection between climate and orbital variations, one would expect ice ages to have occurred at other specific times in the past, which the evidence does not appear to suggest. In the Cretaceous, for example, there seems to have been a seasonless period extending over tens of millions of years.

The biological invasion from space

When the Earth condensed from the primitive nebula, the interior of the Earth melted, and with it perhaps the whole Earth. In the latter case all the organic molecules from the primitive nebula would have decomposed into a mixture of carbon monoxide, methane, hydrogen, nitrogen, ammonia and water. Once the Earth cooled down it acquired a solid crust, and after this, organic material from the remains of the primitive nebula may have been carried to the youthful Earth within meteorites—a sort of parcel post—and influenced chemical evolution. The *possibility* at least existed, for even today meteorites are depositing organic molecules on Earth. In this the principal agents are meteorites of a special kind, known as *carbonaceous chondrites*: they are of the right size not to burn out in the atmosphere, and move slowly enough not to explode on striking the Earth's surface.

There have already been several false alarms regarding organic molecules supposedly found in carbonaceous chondrites. A famous example was the Orgueil Meteorite, which fell in France in 1864. When it was investigated 100 years later, complex organic substances were found, and even cell-like forms that looked like spores. But during the time when the meteorite had been lying around on Earth it could easily have been contaminated by terrestrial influences. 'Probably all we have found is the Earth's thumbprint,' said a critical chemist. In view of this, analyses of this and other old meteorites are no longer regarded as reliable.

But even the sceptics eventually had to concede. The proof fell from the sky at 11 a.m. on 28 September 1968, and landed by the Murchison River in Western Australia. It was a carbonaceous chondrite, whose fragments were collected a few months after the fall: this time the interval was too short for the interior of the newly fallen meteorite to become contaminated, and fortunately it could be analysed in an ultra-sterile laboratory at NASA, which had just been built for the investigation of the Apollo Moon probes. In these sterile operating conditions, the cosmic chemists at NASA's Ames Research Center in California found a variety of *amino acids*, which could only have been enclosed in the meteorite in an extraterrestrial environment, perhaps indeed in its place of origin.

Confusing though the terminology of molecular biology may seem, we need only concern ourselves here with two types of molecules in order to understand how the evolution of life got under way. They are the most important constituents of all life: proteins and nucleic acids. Two kinds of nucleic acid, DNA (deoxyribonucleic acid) and RNA (ribonucleic acid) govern the activities in cells. They owe their name to the fact that they commonly occur in cell nuclei. Proteins comprise the greater part of the fabric of every cell, their complex molecules consisting of long chains composed of amino acids. Exactly 20 amino acids occur in the proteins of terrestrial life, which are therefore built on an 'alphabet' of 20 'letters'.

Nucleic acids, the carriers of genetic information, on the other hand, are composed from only four basic components, the *nucleotides*. Nucleic acids display a more complex structure than the chain-molecules of proteins. They are arranged like two intertwined spiral staircases—the famous 'double helix'. This significant discovery was made in 1953 by the biochemists James Watson and Francis Crick, who received the Nobel Prize in recognition.

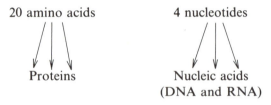

Organic molecules on Earth demonstrate a particular property which marks them out from other molecules. Their components can be joined in two ways, which are mirror images of each other, rotating either to the left or to the right. The molecules of terrestrial life possess without exception a left-handed structure. The fact that only left-handed amino acids are found in all forms of terrestrial life points to a common origin of life; for if we produce organic molecules artificially in the laboratory, left-handed and right-handed molecular variants form in about equal numbers.

Since they were on the lookout for 'alien' right-handed amino acids, biochemists were in a position to distinguish terrestrial from extraterrestrial life.

A research team led by Cyril Ponnamperuma of Maryland University found in the Murchison Meteorite just as many right-handed as left-handed specimens of each amino acid, exactly as would be expected for prebiotic evolution in interstellar clouds. This group also identified right-handed amino acids in old meteorites such as the Orgueil Meteorite.

In 1972 yet another scientific treasure was discovered. Examining the Mighei Meteorite, which fell in the Ukraine in 1889, the Soviet geochemists Alexander P. Vinogradov and Gennady P. Vdovykin of the Vernadsky Institute for Geochemistry and Analytical Chemistry came across a remarkable organic molecule which possessed a double-helix structure like that of DNA.

The molecule in the Mighei Meteorite could not have originated on Earth. Unlike terrestrial DNA it did not show the characteristic left-handed rotation, but was symmetrical. Like other molecules in space, it must have formed on the surface of interstellar dust. The discovery showed how far organic development can go in an interstellar gas cloud, even though radio telescopes have not yet furnished evidence of molecules so complex.

The organic molecule formaldehyde, also found in interstellar gas in 1969 (and since then even in neighbouring galaxies), was also embedded in the Allende Meteorite, which fell in Mexico in 1969. A team from the United States Geological Survey and the Smithsonian Institution in Washington, which had analysed the Allende Meteorite in 1972, tried to estimate the amount of organic material that could have fallen to the Earth in meteorites in early geological times. According to their calculations, a meteorite invasion from the Solar System's cloud of comets may have enriched the seas of the youthful Earth by as much as 100 million tons of amino acids and formaldehyde.

This, of course, does not prove that prebiotic and biological evolution on Earth was actually set in motion by organic material from the Solar Nebula. Analysis of the carbonaceous chondrites nevertheless revived the old panspermia hypothesis ('universal life'). According to this idea microorganisms, driven by the pressure of starlight, were carried into space from the planetary systems of other stars. If they happen to reach a planet with conditions favourable for life, these spores initiate life locally. Svante Arrhenius, the founder of mass action in chemistry, had suggested in 1908 that spores, probably riding on dust grains, could have reached the Solar System in this way. And the physicist Lord Kelvin believed that the first organisms came to us in a comet. Since then both hypotheses have become hopeless outsiders. Calculations made in the early 1970s showed that any known type of radiation-resistant spore would have been subjected to much too great a dose of radiation to survive the long journey between two stars. Nor are the prospects good for Kelvin's comet idea: the probability that a large-enough body could leave one solar system and land on the planet of another is so small that in almost 5000 million years a meteorite from outside the Solar System has probably never reached the Earth. This makes the 'infection theories' of the nineteenth century appear most implausible.

In 1973 Francis Crick and Leslie Orgel started another attempt to revive the panspermia hypothesis.[27] An extraterrestrial civilization, which perhaps faces the danger of dying out, or wishes to spread its own form of life, might send a spaceship to another planet with a cargo of microbes. 'The spaceship would carry large samples of a number of microorganisms, each having different but simple nutritional requirements, for example blue-green algae, which could grow on CO_2 and water in "sunlight". A payload of 1000 kg might be made up of 10 samples each containing 10^{16} microorganisms, or of 100 samples each of 10^{15} organisms.' Crick and Orgel maintain in support of their 'directed panspermia' that all types of terrestrial life have the same biochemical basis for their reproduction—the genetic code which is universal on Earth. 'It is a little surprising that organisms with somewhat different codes do not coexist,' they comment. Perhaps terrestrial life originated from a single microorganism from another planet.

Although this hypothesis can never be completely ruled out, it does not seem very probable. (We should also mention the difficulties that exist in general for space probes and space travel, as discussed in Part III.) And as long as the origin of life can be explained entirely by processes on Earth and within the Solar System, even a modified panspermia hypothesis need not be introduced.

The primeval soup and evolutionary reactors

Before the influence of meteorites on the evolution of life had again begun to be seriously considered, biochemists were naturally principally concerned with the question of how organic molecules could have originated on Earth in primitive conditions. At all events, there was no shortage of energy sources capable of directing a chemical synthesis. The most energy was provided by the Sun, primarily through ultraviolet light. The electric discharges of thunderstorms also filled the primitive atmosphere; in addition, energy for chemical reactions was provided by the shock waves produced by meteors in the atmosphere, as well as by the radioactivity of the Earth's crust and the heat output of volcanoes. Many laboratory experiments have been performed to simulate these primordial conditions *in vitro*: these are first-generation evolutionary reactors.

The breakthrough was achieved in 1953 by the chemist Stanley Miller of the University of Chicago. He produced in a glass vessel conditions like those in the

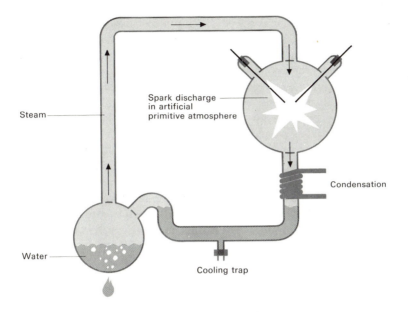

Figure 9 Miller's first-generation evolutionary reactor. The reactor simulates conditions on the Earth about 4000 million years ago. The steam given off by the boiling water in the flask causes an artificial primitive atmosphere of hydrogen, methane and ammonia to circulate past the tungsten electrodes, where spark discharges strike through the gas mixture. The water in the flask (imitating the primitive oceans) fills with condensing amino acids—the simplest building blocks of life.

secondary primitive atmosphere, a gaseous mixture of methane, ammonia, hydrogen and water vapour. Through this mixture Miller passed electric discharges simulating the lightning flashes of primeval thunderstorms. The discharge products were condensed in a tube (see Figure 9). Miller ran the experiment for a week, then he pumped off the gases and examined the liquid. He had produced the first artificial organic 'primeval soup'—in an experiment which had in fact only repeated the necessary preliminaries for the evolution of life: a purely chemical synthesis.

Miller had actually expected a rich mixture of many different, but very simple, molecular compounds; however, relatively few types of molecule were produced. Furthermore, 'the major products were not themselves a random selection of organic compounds but included a surprising number of substances that occur in living organisms'.[28]

The most abundant products of the experiment were the amino acids glycine, alanine and aspartic acid, which are among the structural units of proteins (see Table 3). Evolutionary theorists appeared to be on the right track. But, as a further surprise, the artificial primeval soup also contained some amino acids which do *not* occur in terrestrial proteins, which are built from only 20 different amino acids. Thus a clear selection from among many possible amino acids must have taken place in the evolutionary process.

Since Miller's evolutionary experiment, scientists have repeated the investigation in different conditions. Instead of electric sparks they have used light or bombardment by high-energy particles such as the solar wind carries to the Earth. They have also passed shock waves through the simulated primitive atmosphere, like those produced by meteors as they fall to the Earth. In every case amino acids were formed in the artificial primeval soup. Heating by shock waves even turned out to be the most effective way of converting ammonia into amino acids. Nevertheless, the concentration remained too low for it to have led to further reactions between biomolecules in the open oceans. Before this could happen, the primeval soup must have 'thickened'. It is generally supposed that this took place in ponds or other small bodies of water at the edges of seas. The prebiotic synthesis of some amino acids thus seems to be explained; but biochemists have still to show how *all* the 20 amino acids were formed by 'natural' means from their atomic building blocks. On the other hand, it was definitely shown that *no* amino acids were formed if the simulated primitive atmosphere was 'oxidized', that is oxygen-enriched, and above all if it contained no hydrogen. Amino acids only survived if the substances involved in the chemical reactions of the primitive atmosphere were 'reducing', that is made proportionally weaker in oxygen rather than stronger. This dependence on the reducing nature of the primordial atmosphere has led to criticism of the 'primeval soup' concept, but more on this later (see page 43).

But the amino acids that arrived on Earth with the Allende and Murchison meteorites certainly did not originate in an aqueous broth. In interstellar gas, at the edge of a primitive Solar Nebula and in the absence of energy from ultraviolet light or lightning discharges, the reaction must have taken a different course. After the radioastronomers at Green Bank discovered ammonia and formaldehyde in interstellar clouds in 1969, two biochemists got down to work. In 1970, in a laboratory at Miami University, Sidney Fox and Charles Windsor heated ammonia and formaldehyde together and obtained several amino acids, especially alanine and glycine, which Miller's evolutionary reactor had also produced in the greatest abundance.

Biochemists have not been so fortunate with the prebiotic synthesis of those important building blocks of life, the four nucleotides for nucleic acids. But let us first take a closer look at the nucleotides. A nucleotide consists of three components: a sugar, a phosphate group and one of four nucleic acid bases. The bases determine, so to speak, the character of the nucleotides, and belong to either the *pyrimidine* or the *purine* group. Pyrimidines are ring molecules consisting of six atoms; they form a six-sided ring of nitrogen and carbon atoms. In the nucleic acid bases of the purine group, a five-sided ring is attached to the six-sided pyrimidine ring. The nucleic acid DNA contains two pyrimidines called *cytosine* (C) and *thymine* (T), and two purines, *adenine* (A) and *guanine* (G). The same building blocks form the structure of RNA, with one exception: there are adenine, guanine and cytosine, but the pyrimidine thymine is replaced by the pyrimidine *uracil* (U). We thus have in either case four building blocks, the 'letters of the genetic alphabet':

DNA		RNA	
Adenine nucleotide	= A	Adenine nucleotide	= A
Cytosine nucleotide	= C	Cytosine nucleotide	= C
Guanine nucleotide	= G	Guanine nucleotide	= G
Thymine nucleotide	= T	Uracil nucleotide	= U

DNA and RNA are filamentary molecules, consisting of a string of their respective letters. They are chain molecules, which can be several million units in length. But the sequence of the letters is by no means random. Each group of three letters constitutes a functional unit: a selection of four letters gives $4^3 = 64$ different triplets—although there is not an amino acid for every combination. In the cell the triplets play a fundamental directive role in the formation of proteins from the 20 amino acids. More precisely, each triplet determines the insertion of a particular amino acid in the chain molecule of a protein. For example, in RNA the triplet GCU carries the information for the insertion of the amino acid alanine. Table 3 gives a summary of which amino acids are inserted by the possible 64 triplets. As each triplet represents a particular information unit, a sort of telephone number or code word in the language of macromolecules, biochemists also call it a *codon*. But, in effect, only 20 of the 64 codons are needed for the 20 different 'natural' amino acids. What are the meanings of the other codons? Research has shown that several codons can be responsible for the same amino acid: this can also be seen in the table. For example, the codons GCU, GCC, GCA and GCG all code for alanine. The *genetic code*, then, contains redundancy. Some codons fulfil the function of punctuation marks; they separate words and sentences like stops and commas. Because of the redundancy of the genetic code, 61 codons in fact find a use in coding for amino acids, and the remaining 3 are used as punctuation marks, ending chains.

The remaining components of nucleic acids, the sugars ribose (in RNA) and deoxyribose (in DNA), can form under prebiotic conditions from formaldehyde, one of the first substances to form in Miller's evolutionary reactor. The prebiotic synthesis of ribose was first achieved experimentally in the nineteenth century by a Russian scientist, A. M. Butlerov, who, by shaking formaldehyde with calcium carbonate, produced a mixture of sugars that included the pentose ribose.

Of course, there have also been attempts to produce nucleotides in primitive conditions using the recipe of the Miller experiment. One organic base, the purine

Table 3 The genetic code

First position	Second position				Third position
	U	C	A	G	
U	Phenylalanine	Serine	Tyrosine	Cysteine	U
	Phenylalanine	Serine	Tyrosine	Cysteine	C
	Leucine	Serine	End chain	End chain	A
	Leucine	Serine	End chain	Tryptophan	G
C	Leucine	Proline	Histidine	Arginine	U
	Leucine	Proline	Histidine	Arginine	C
	Leucine	Proline	Glutamine	Arginine	A
	Leucine(?)	Proline	Glutamine	Arginine	G
A	Isoleucine	Threonine	Asparagine	Serine	U
	Isoleucine	Threonine	Asparagine	Serine	C
	Isoleucine	Threonine	Lysine	Arginine	A
	Methionine	Threonine	Lysine	Arginine	G
G	Valine	Alanine	Aspartate	Glycine	U
	Valine	Alanine	Aspartate	Glycine	C
	Valine	Alanine	Glutamate	Glycine	A
	Valine	Alanine	Glutamate	Glycine	G

Three nucleotides selected from the four 'letters' U (uracil), C (cytosine), A (adenine) and G (guanine) give the command for the incorporation of a particular acid into a protein molecule. In the scheme here, the triplet is read off as follows: take the first letter from the left-hand column, the second from the top row and the third from the right-hand column. The high redundancy of the third letter can be clearly seen. Several triplets (also called *codons* because of their directive function) may stand for the same amino acid; three give an end chain—a sort of 'full stop'.

adenine, was synthesized in the laboratory as long ago as 1961, in an experiment carried out by the biochemist J. Oró. He dissolved the more simple molecule of hydrogen cyanide (HCN) in water and exposed it to energetic radiation in an artificial primitive atmosphere. Large quantities of hydrogen cyanide had already been produced in Miller's evolutionary reactor, and it has also been observed since 1974 as an interstellar molecule, for example in Comet Kohoutek. It thus seems that interstellar HCN is also a direct forerunner of the organic base adenine. But adenine could equally well have originated on the Earth's surface as a result of the Sun's energy-rich radiation. The other purine, guanine, has also been successfully synthesized by several routes, for instance from cyanogen (CN). But for the other nucleic acid bases, the pyrimidines, the experiments with evolutionary reactors in primitive conditions have failed to some extent—so far, the only pyrimidines that have been synthesized are those which do not occur on Earth. This research problem remains to be tackled. Pyrimidines not represented in terrestrial life-forms were found in the Murchison, Allende and Murray meteorites.

These gaps in the scheme of chemical evolution on an already cooled Earth show that some of our ideas may have to be revised. The gaps may not indicate clumsiness on the part of the experimenters so much as profound deficiencies in the evolutionary model.

Nor has the last word been said, it seems, on 'the primeval soup as the source of life'. In 1977 Fred Hoyle and Chandra Wickramasinghe published a paper in which they dispute that there ever was an amino-acid-rich primordial ocean.[29] Indeed, experiments of the sort used by Miller had demonstrated that no amino acids would form in an oxidizing primitive atmosphere. This took place only in a reducing atmosphere—one rich in hydrogen. Here we encounter a problem. Given a reducing primitive atmosphere and a corresponding 'primordial soup', the oldest rock samples should show large deposits rich in nitrogen and carbon. But geologists have not found such deposits. Hoyle and Wickramasinghe regard this as an indication that there may not have been any primeval soup at all. A possible way out is that there was a primeval soup for a relatively short time, which came to an end before the first sediments were deposited, but lasted just long enough to produce the necessary amino acids and to set going the subsequent evolution of life.

Hoyle and Wickramasinghe believe that comets and meteorites carried to the Earth the organic material which started off this evolution. As the discoveries of such meteorites as the Murchison and Allende show, the Earth is still being bombarded today by extraterrestrial organic molecules. The two researchers therefore pose the question: 'If a cometary impact led to the start of life, ... would subsequent arrivals of cometary material carry biological, or prebiological material which might affect terrestrial biology?' They conclude: 'The boldest answer must be yes; that is to say, extraterrestrial biological invasions never stopped and continue today.'[29] Not unlike the advocates of the panspermia theory, they think that this invasion even involves, among other things, viruses and bacteria. In support of their thesis, Hoyle and Wickramasinghe have drawn attention to several historical epidemics and pestilences, one of which, for example, claimed 30 million lives in 1918/19. They also mention the abrupt appearance of records of cold and influenza epidemics in the fifteenth and seventeenth centuries, as well as earlier plagues which do not have easily recognizable present-day counterparts.

Meteorites may certainly have acted as a catalyst at the origin of life, but more research is needed to clarify the question. It is virtually impossible, however, that the biosphere of the Earth could still be influenced today from without, owing to the universal nature of the genetic language (code) of the molecules of heredity. As will be explained at greater length in the next section, there are in principle many biochemical possibilities, many different codes, by which macromolecules can exchange information among themselves. *Which* code succeeds is probably a matter of chance, but the successful code will then suppress all the alternatives. In protein synthesis all terrestrial life speaks the same genetic language. On other planets or in dense interstellar clouds a given code, also determined by chance, will likewise succeed, but it is most unlikely that this code will be exactly the same as that selected on Earth. Extraterrestrial viruses and bacteria, with another code, would as a result be quite unable to establish biological communication with their terrestrial counterparts. At best they would develop among themselves, but biologically separate from terrestrial life-forms. And in that case we would have been bound to find life-forms with other genetic codes in the terrestrial biosphere; but so far *not one* living thing has been found to use a different code. And so long as no evidence to the contrary is known (of course not *every* living thing has been investigated) we can regard the hypothesis of 'epidemic disease from space' as unproven.

Hypercycles and the rise of macromolecules

Although the prebiotic synthesis of some organic molecules still presents an unsolved problem, the second stage of evolution, after chemical evolution, is today the focus of microbiology: *self-organization of macromolecules*. This stage constitutes the actual boundary between the living and the non-living. But the transition is fluid: viruses, the classical 'hybrids' of the two worlds, are on the one hand parasites and multiply within the host cell at its expense. But without a host cell a virus may behave like a particle of non-organic material. It can, for example, integrate into a crystalline structure. Where we should draw the frontiers for the origin of life was first postulated in 1924 by the Soviet biologist A. I. Oparin, who made it dependent on satisfying the following three criteria:[30]

 • Self-reproduction (replication)—to preserve the organization, which is constantly exposed to destruction by external conditions.
 • Mutability—for the development of new information which could lead to the origin of new species, according to the Darwinian selection principle.
 • Metabolism—energy exchange (by chemical interaction) with the environment, in order to maintain the non-equilibrium condition that constitutes life.

These criteria are necessary conditions for molecular self-organization—'the nucleic acids as the seat of the Legislature, and the proteins as bearers of the Executive', as Manfred Eigen of the Max Planck Institute for Biophysical Chemistry in Göttingen has described it.[31]

On the nucleic acids as information banks falls the task of replication. Each chain molecule of DNA—coiled around itself in the now-familiar 'double helix' of Crick and Watson— consists of two strands of nucleic acid bases ('letters'), with each base in the strand paired against a quite specific partner in the other strand. For DNA the base pairs are $A \leftrightarrows T$ and $C \leftrightarrows G$, and for RNA $A \leftrightarrows U$ and $C \leftrightarrows G$. (The double arrow indicates that in each case the base represented by the letter on the left is held by hydrogen bonding to that on the right, and conversely the letter on the right attaches itself to that on the left.) In the replication process, the two intertwined strands of the molecule separate, just like a zip-fastener, so that the pairs of bases are successively parted. Each individual 'letter' then serves as a template for its complementary 'letter', such that freshly formed nucleotides attach in sequence to the exposed string. In this way, each strand of the double-stranded nucleic acid molecule acquires a complementary strand, forming a complete new DNA molecule, and the original molecule becomes duplicated.

An example is the bacillus *Escherichia coli*. Its DNA contains about four million molecular 'letters'—this corresponds to a ponderous tome with 1000 pages. Guided by enzymes, it can replicate completely within 20 minutes. In this interval four million 'letters' are read off, and all the steps of the reaction—which often require only a millionth of a second each—are carried out.

But how did these macromolecules originate, how did the meticulous microclockwork of 'legislative' and 'executive' reactions come about? It was surely not due to chance, which could not have produced every configuration of molecular building blocks even in the most primitive stages of organization. Even a relatively simple protein, with 100 amino acid components, and with 20 candidate amino acids for each position in the molecule, would be but one example out of 20^{100} possible combinations. But the whole Universe contains no more than 10^{81} particles! Only an evolutionary adaption process could have prevailed against this statistical improbability. The biologically active macro-molecules must have adapted step by step to the conditions of their surroundings and to their functions. And this will have taken place long before the development of the most primitive cells. 'This means that Darwin's principle of natural selection is already valid in the molecular domain.'[31] Thus the molecular species were already in competition, as bacteria, plants, fishes or mammals were later. But this is not to say that the Darwinian process was in any sense 'inborn' in the molecules through selection and evolution. It is rather that natural selection is an inevitable characteristic of a population consisting of a variety of reproducing organisms which are subject to external constraints, such, for example, as a finite, limited and constant supply of food. Darwinian living creatures, to whom the theory of natural selection applies, have to multiply, live for a limited time, and compete with one another as they develop. This is exactly what macromolecules do—they reproduce themselves, demolish themselves chemically after a certain time, and compete for structural materials; and thus the fastest-growing molecular chains acquire an advantage in the selection process.

Exact conditions for the selection and evolution of molecules can be derived with the aid of the theory of non-equilibrium states, for which I. Prigogine received the Nobel Prize in 1977. No life can exist in (thermodynamic) equilibrium: life is a non-equilibrium state, which can only be sustained by a steady flow of energy. Nevertheless, the life-form must at times be stable, that is, stationary. Just such biologically stable states are the result of selection: the selected species, whether macromolecule or living thing, has prevailed—this is the stable aspect—over all competitors less well adapted to surrounding conditions; but it is conquered by new species which reproduce better in the given circumstances.

Propagation and mutability—the power to produce mutations while propagating—are necessarily closely linked. Mutations occur when copying errors intrude during the replication process. (Mutations caused by natural radioactivity and cosmic rays are statistically unimportant.) Without copying errors (that is with perfect replication) no mutations would occur, and a race could not adapt itself by selection to a changing environment. On the other hand, there must not be too many errors in the copying, as this would destroy too much of the very information which is meant to be inherited. The errors would be faithfully copied in further replication, and after several generations the species would have degenerated. The genetic information 'leaks away'. Therefore the capacity of the molecular information bank has a definite ceiling: with a fixed error rate (or a set level of copying accuracy), the nucleic acid holding the information cannot grow to too great a length. If, however, it is made longer, the accuracy of copying must improve so that the percentage error does not increase. With a constant rate of error, on the other hand, the longer the molecule, the more errors will be made. As a consequence, the information will be lost after a few generations. But if the error rate is too small, the evolutionary process may be too slow because too few mutations are produced.

There is an important difference between the selection processes of macro-molecules and biological life-forms, which lies in the nature of the contest. Whereas cells, plants and animals will tolerate the presence of many species around them, the struggle of molecules in given conditions leads to an all-or-none decision. The reason for these differing competition 'rules' can be found in the amount of information necessary for a reproductive operation. In order to produce a copy of itself (*autocatalysis*), a molecule and the associated translation machinery require a *minimum* amount of information, just as a recipe must contain a minimum number of words to make the meal a success. Manfred Eigen and his colleague Peter Schuster in Vienna, in their investigations of parasitic viruses (the simplest reproducing molecules), have established the smallest amount of information which might be found in the complexity of their structure.[32] Mathematical analysis of the experiments showed that the amount of information stored in the viruses is greater than can be reproduced with sufficient accuracy in a simple autocatalytic system. The constant rate of copying errors in such a system, in attempting to carry over too much information at once, would destroy the content by introducing too many errors. The information would 'leak away'. The German science writer Thomas von Randow outlines the resulting dilemma as follows: 'This maximum information content which such a system *can* meaningfully reproduce is still considerably smaller than the minimum information it *must* reproduce if the whole mechanism is to continue to be self-reproducing.'[33]

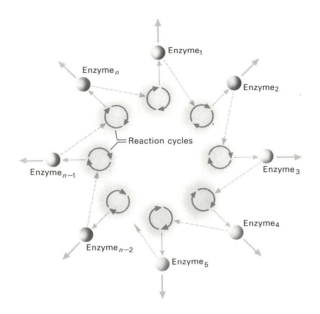

Figure 10 A particular type of co-operation between biological molecules is illustrated by the Eigen hypercycle. The units of a hypercycle are themselves reaction cycles, in which certain molecules produce copies of themselves. Each time they also produce an enzyme, which acts upon a further reaction cycle. When the last enzyme produced (enzyme$_n$) acts upon the first reaction cycle, the cycle of reaction cycles (the hypercycle) is completed. At this point, the whole reproductive system becomes self-reinforcing. (After ref. 32.)

According to Eigen it could have come about in the following way. The different autocatalytic systems began to co-operate with one another, so that while each individual one only had to reproduce the amount of information it was able to handle, the different systems acting together reproduced the information necessary for reproduction. From an initially rudimentary and more or less accidental co-operation among individual replicating systems there developed a clearly defined copying process. For this to win through against competing copying processes, the whole co-operative had then to become stabilized, for it was only as a unit that it could complete its task. Eigen has given the type of organization by which the coalition of replicating components will most easily acquire a *common* selective advantage the name of *hypercycle* (see Figure 10). He sees the functions of the individual partners as a cyclical relationship: 'The translation product of nucleic acid A favours the reproduction of nucleic acid B. Its respective translation product operates on C, and this variety of reciprocal relationship continues until finally the translation product of nucleic acid X feeds back to A and thus completes the cycle. In this network of reactions, each nucleic acid molecule A, B, C . . . X represents a reproductive cycle of its own, which is a closed system by virtue of its positive–negative complementarity. But it is only the higher or "hypercyclic" association which provides the common selective advantage,' writes Manfred Eigen.[31] The hypercycle supplies a form of organization by which the amount of information necessary for replication can be stored and processed with sufficient precision.

This network is self-stabilizing, and brought about a general selective advantage, which finally exterminated all competing organizations. The reason is that the molecules grow in number 'hyperbolically' through their indirect self-reinforcement. While exponential growth—like Darwinian selection of cells and higher life-forms—can still tolerate a large number of species side by side, non-linear selection leads to an 'all-or-nothing' struggle. This most probably explains two fundamental properties of terrestrial life, which we have already mentioned without comment: the universality of the genetic code on Earth (see 'The biological invasion from space', page 35) and the fact that the naturally occurring amino acids in terrestrial life-forms are all without exception of the left-rotating (left-handed) type. It is likely that the primeval soup contained about equal numbers of right-handed and left-handed amino acids, like the artificial primeval soup of Miller's evolutionary reactor. Then certain fluctuations led to the left-handed amino acids becoming rather commoner than the competing mirror images. Non-linear coupling in a hypercycle turned this head start into a large lead. The lost ground could not be recovered, and the competitors paid the price: they died out, evidently at a time before the first fossil remains were deposited. (Possibly the oldest microfossils and organic compounds, with an age of about 3400 million years, are those discovered at Pilbara in Western Australia.)

Exactly the same must have happened with the genetic codes. Macromolecules which had developed a particular code occasionally became more common as a result of statistical fluctuations. In a phase of hypercylic co-operation, this was enough to win them the battle. Competitors with other codes died out, and only one survived, a 'frozen accident'. To Eigen, this universality indicates 'a shared, but not necessarily unique original event'. Had there not been such a radical all-or-none competition, but had several initial events been permitted to carry through, various genetic codes and different rotating schemes would be found today in terrestrial life-forms. But these have not been found. We know that there was no going back on the decision concerning the fundamental structure of life; at no later time has another code taken over.

Eigen and Schuster close their article of November 1977 with the following notable statement: 'If we are asked, "What is particular to hypercycles?", our answer is, "They are the analogue of Darwinian systems at the next higher level of organization."' Darwinian behavior was recognized to be the basis of generation of information. Its prerequisite is [the] integration of self-repro-

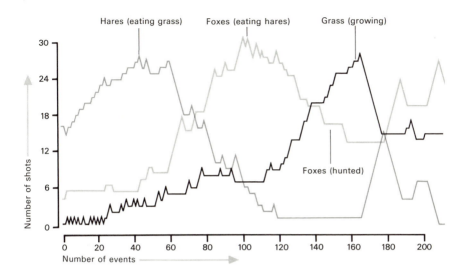

Figure 11 Grass, hares, foxes and huntsmen: game-play model of a hypercycle. The changes in the different populations with time are clearly reflected in waves of varying phases. The individual components—grass, hares and foxes—are distinguished by lines in different shades. (After ref. 34.)

ductive symbols into self-reproductive units which are able to stabilize themselves against the accumulation of errors. The same requirement holds for the integration of self-reproductive and selectively stable units into the next higher form of organization, in order to yield again selectively stable behavior. Only the cyclic linkage . . . is able to achieve this goal.'[32]

Eigen and his co-worker Ruthild Winkler have developed models for hypercycles using games theory. In their book *Das Spiel* they deal with models which they have played through with a computer; for example, 'Grass, Hares and Foxes' (see Figure 11). These protagonists co-operate and compete in the pattern of a hypercycle. '1. First, grass grows. 2. The grass is eaten by hares and the number of hares increases. 3. Foxes eat hares, and the number of foxes increases. 4. Foxes are hunted.'[34] While the hares are still being eaten, the grass can already grow again. As soon as the foxes disappear the hypercycle is closed. Experimental confirmation of a hypercyclic self-organization in macromolecules is still awaited, but significant partial results have already been obtained in the 1970s.[32,35,36]

The synthesis of enzymes from simulated primitive conditions presented greater difficulties than that of amino acids until quite recently. But according to a colleague of Eigen, Bernd Küppers, following evolutionary experiments of the second generation (see Figure 12) it is regarded as 'established that under

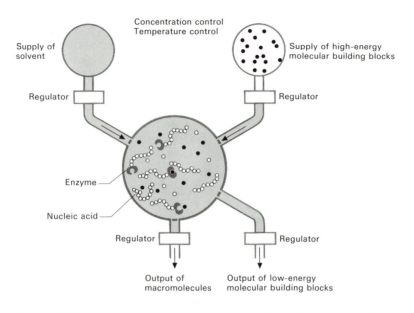

Figure 12 Second-generation evolutionary reactor. Controlled amounts of energy-rich molecular building-blocks (●)—nucleoside triphosphates of nucleotides A, U, G and C—flow into the reactor along with controlled quantities of a solvent. Synthesized macromolecules (∞∞ = RNA) and low-energy breakdown products (○) flow from the reactor. The system also contains an enzyme—in an accurately measured concentration—which links up the nucleic acid building-blocks in a chain. The building-materials are radioactively labelled to allow continuous control and analysis of the molecular composition of the contents of the reactor. (After ref. 37.)

appropriate conditions, i.e. in a reducing atmosphere with various energy sources, both nucleic acids and proteins can originate independently of each other. Thus the existence of biological macromolecules is sufficiently explained by our present concepts in chemistry and physics.'[37] For many biochemists there is no longer any real doubt: macromolecules first amplify their information content by simple replication and mutation, impelled by the inexorable course of Darwinian selection. The origin of life was a necessary consequence—and no 'cosmic accident'—once the physical and chemical conditions allowed an automatic increase in the molecular concentration on the pattern of a hypercycle (see Figure 13). Just how correct these assumptions are has thus far been little explored, but a primeval soup in a pond at the edge of the ocean seems to have been the right place.

It seems to me, however, that a cosmic accident in a way still remains possible, even though neither so improbable nor of the same kind as that advocated by Jacques Monod and others for the origin of life. As the discussions elsewhere

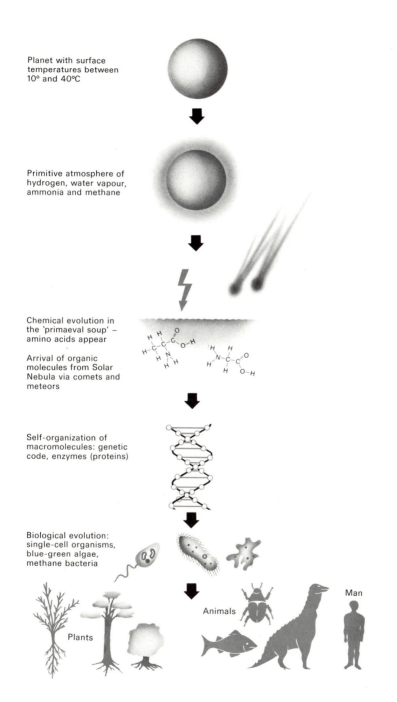

Planet with surface temperatures between 10° and 40°C

Primitive atmosphere of hydrogen, water vapour, ammonia and methane

Chemical evolution in the 'primaeval soup' – amino acids appear

Arrival of organic molecules from Solar Nebula via comets and meteors

Self-organization of macromolecules: genetic code, enzymes (proteins)

Biological evolution: single-cell organisms, blue-green algae, methane bacteria

Plants

Animals

Man

Figure 13 The origin of life on Earth.

indicate (see 'The Earth's atmosphere: the minimum conditions for life', page 27), accident is part and parcel of the external circumstances of planets and stars. If the ecospheres of the stars are in fact very small, a hypercycle between nucleic acids and proteins will far less often get under way. On Earth, life came into being 4000 million years ago as a matter of course. But planets like the Earth, which moreover move just in the narrow ecospheres of stars like the Sun, are probably none too common. And even if genetic molecules, bacteria and unicellular organisms arise on a few million planets, how far will more advanced life-forms develop, what degree of complexity will they attain? Can evolution stop and stagnate 'half-way', or will it inevitably develop as far as mankind and beyond?

The genetic future of mankind

It is popular to speculate whether the future evolution of mankind may, perhaps by the genetic route, make another 'leap forward', towards greater complexity— possibly influenced by the achievements of genetic engineering, which, for instance, von Däniken's gods are supposed to practise. But it seems that the genetic complexity attained by human cells represents an upper limit and that our genes already process the maximum amount of biochemical information. The key to the proof of this assertion lies in the frequency of errors made by macromolecules during replication. Compare, for instance, the error rates in the genetic processes of various life-forms (Table 4). The simplest biological systems, such as viruses, produce one incorrect copy per thousand ($1:10^3$) correct readings of genetic symbols. The cause of these transcription errors is the perpetual motion of the molecules—their thermal energy. This produces 'noise' disturbing the short copying process between two molecules, which lasts only a millionth of a second, and impairs the accuracy of the reading. If the proportion of error in viruses were greater than $1:10^3$, every copy of a chain of 1000 'letters' would contain a defect which, by perpetuation and accumulation, would result in degeneration and the loss of genetic information within only a few generations.

But bacteria, which require a reliable duplication of millions of individual symbols, can no longer permit themselves an error rate of more than one in a million ($1:10^6$). Manfred Eigen and Peter Schuster have worked with *Escherichia coli*, whose nucleic acid contains four million nucleotide units. The two investigators discovered that for *E. coli* the ratio of molecule length and error rate lies near the calculated optimum at which evolution can proceed at the fastest rate. Above this threshold all the information previously accumulated by evolution is lost. Finally, in human genes thousands of millions of symbols are copied. Correspondingly, an error is made only once in several thousand million readings. 'An increase in information capacity could thus only be achieved if

Table 4 Transcription errors in different life-forms

	'Letters' (nucleotides) maximum	Repair system	Error rate
Viruses	10^4	None	$1:10^3$
Bacteria	10^7	Enzymes	$1:10^6$
Cells	10^{10}	Genetic recombination	$1:10^9$

there was a constant improvement in the accuracy of reproduction,' comments Eigen. This increase and the reduction of the error rate by a factor of 10^3 in the transition from virus to bacterium required mechanisms of a new kind—biochemical repair systems with enzymes specially developed for the purpose. These effect repairs by checking the copy of a molecule, discovering the damaged spots and surrounding each one like a sleeve. How about man? Only a further device could again reduce the error rate to a thousandth. And it took 2000 to 3000 million years before evolution had produced a still more sophisticated mechanism: genetic recombination, the basis of sexual reproduction (see Figure 14).

Can the information capacity of cells surpass that found in mankind, whether by evolution or by artificial operations? Could there be a still more sophisticated repair system capable of exceeding even the precision of human reproduction?

Perhaps Nature has already played her best and final trump card with sexual reproduction: the upper limit of reproductive precision seems to have been reached. For even the best repair system cannot completely suppress the statistical copying errors introduced by molecular thermal motions if the number of readings grows too great. Although man may one day be able to make better use of his brain, he would do better not to anticipate a still more complex genetic future. This upper limit for genetic complexity applies equally to extraterrestrial organic life-forms, however different they may be in genetic structure and appearance. According to our present understanding of the phenomenon of life, genetic complexity in extraterrestrial civilizations will not exceed that of man. Some molecular biologists suspect that, on the contrary, we can expect genetic degeneration in the future.

The future prospects of genetic engineering, too, are often over-estimated. At all events, the idea that we may some day be working synthetically with macromolecules consisting, like human genes, of thousands of millions of molecular symbols, is consigned by biochemists to the realm of fairy stories. A biologically active macromolecule, whether protein or nucleic acid, operates not only by the sum of the individual functions of its components, but also by their grouping in larger associations within the macromolecule. It is like a gigantic

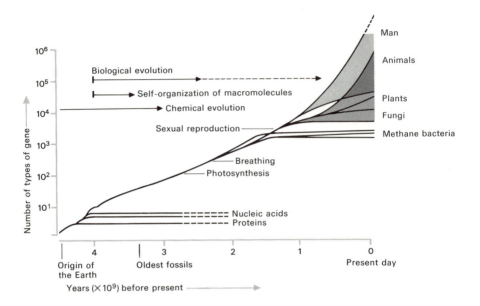

Figure 14 Level of organization of evolution, measured by number of types of genes.

mathematical formula with a hierarchy of brackets, multiplication signs and cross-relations, which operate at a distance across many neighbouring units within the molecule. Even for bacteria with strings of molecules of 'only' a few million units, it is unlikely to be possible in the foreseeable future to synthesize in a retort such a thing as a molecule which will perform a function precisely laid down in advance. How far genetic engineering can go, through operations on the genetic material already available in human cells, is still beyond the power of prophecy. In this regard, the speculations of the Princeton physicist Freeman J. Dyson should be regarded critically; he envisages for the future a complex microbe technology, which will not only eliminate environmental pollution but also make the Solar System's asteroids and comets habitable (see Part II: 'The astroengineers of supercivilizations', page 200).

But this does not rule out small successes (and potentially dangerous failures). These are already being achieved by directed operations on quite complex biomolecules. In May 1974 genes from a rat cell were 'spliced' into a bacterium in an attempt to produce insulin.[38] The genetic operation, known as cloning, was performed by the Californian biochemist Howard Goodman with the bacillus *E. coli*. First the rat genes, which contain the information for insulin production, were copied, and then the copy, directed by enzymes, was inserted in the genetic molecules of the bacillus. The manipulated bacteria replicated the foreign gene material when they multiplied, and thus passed on the transplanted insulin code.

Goodman and his colleagues believe that the prototype of a new insulin producer can soon be artificially produced, which in time can supplement the production of insulin extracted from the pancreatic juice of slaughtered cattle. An important step was indeed taken with the inheriting of elements of the insulin gene, but the altered bacteria have hitherto produced no insulin. They do not 'recognize' the intruding genetic message as a command actually to produce insulin. This problem must be solved if Goodman's high hopes are to be fulfilled. 'To reach this goal, [the implanted insulin gene] must be carefully coupled to bacterial recognition mechanisms. We may be certain that this is possible, though difficult. Positive experiments with natural nucleic acids are yet to be performed,' notes Peter Hans Hofschneider of the Max Planck Institute for Biochemistry at Martinsried near Munich.[39]

In plant breeding, great hopes are pinned on genetic operations as a way to introduce changes. Hitherto, the yields of useful plants have been increased by nitrogen fertilization and modern agricultural methods. Hopes of progress are pinned onto attempts to induce plants to obtain nitrogen not by the roundabout method of fertilization, but directly from the air. There is one family of plants which does this: the Leguminosae. The bacteria in their root nodules render the atmospheric nitrogen directly available for the nutrition of the plants. An aim of plant geneticists is to transfer this system to other plant families, but it has hitherto proved vain. Another speculation from the witches' cauldron of genetics concerns the process of ageing. Many scientists believe that highly developed extraterrestrial civilizations will have solved the problem of ageing. This optimism is altogether understandable, for the history of the extension of our lifetimes has run an astonishing course hitherto. At the time of Christ a member of Roman society died on the average at the age of 23. In 1850, life expectancy had risen to 40 years, and today it stands as high as 70. It took almost 2000 years, therefore, for the life expectancy of a newborn child to rise from 23 to 40 years. Thereafter, life expectancy rose within 120 years from 40 to 70 years. This result was not, however, a true 'extension' of life. It was principally brought about by an increase in hygiene and the reduction of infant mortality.

The question is how long this development can continue, and what other diseases men would die of at a yet more advanced age. The possibilities of extending life by hygienic methods and the control of infectious diseases seem to be practically exhausted. People die today principally from diseases of the heart and vascular system (heart attacks, arteriosclerosis and suchlike) and cancer; medical research is already working on the reduction of mortality from these causes. According to one estimate, a 50 per cent reduction in the number of deaths from heart and vascular disorders may be expected to extend the life expectancy of 65-year-olds by around 4 years. Hopes of extending life therefore often rest on directed manipulation of the cell functions. However, according to our present knowledge, it will only be possible to achieve a further doubling of the present life expectancy. This is essentially because human cells can regenerate themselves by division at the most 50 times. With a greater number of divisions, biochemical processes within the cell are less and less well maintained, until finally the cell breaks down. Several years ago the biologist Leonard Hayflick, from the Oakland Medical Center In California, showed that 50 divisions is the limit, even if (as has been done with embryonic cells) they are frozen for some years after the twentieth division: these, after being thawed out, divided 30 times more. The command to divide is given by the DNA in the nucleus. The evidence for this was presented by Hayflick at a conference in December 1977.[40] He isolated the nuclei of an old and a young cell, and inserted the young nucleus in the prepared old cell. The result was that the rejuvenated cell divided itself correspondingly more times on the command of the young control centre.

The ageing process has two causes. One is the daily damage suffered by the genetic molecules. In a young cell these defects are very rapidly taken care of: repair enzymes exchange the damaged sections with fresh ones. But protein fragments remain as waste products of the repair, and they must be dismantled by the cell. 'Repair shop and refuse disposal' describes the business of cell maintenance. Laboratory research has shown that the repair efficiency falls off after the thirty-fifth cell division, and that after the fortieth division refuse disposal falls down on the job by 30 to 70 per cent. Finally, after the forty-fifth division damaged genes and an excess of waste impede the cell's functions, and after the fiftieth division at most the cell dies of old age. There are substances, especially that known as Centrophenoxin, which will remove the accumulating injurious material during the declining stages, as the Stuttgart biologist Klaus Bayreuther has successfully demonstrated in his tests on the cells of rats. He said of his work: 'We can now increase the life span of rats and mice by 30 to 40 per cent.'[40] Other materials are capable of suppressing the production of waste products.

How far these results can be tested on man depends on the safety precautions and controls applied to future genetic research. One day, perhaps, people 150 years old will no longer strike us as unusual. But whether it is desirable to grow so old while the body is kept young—for the brain cells do not renew themselves—is quite another question. And in any case, this has nothing to do with the goal of quasi-immortality.

Alternative life-forms: crystal beings and methane breathers?

One important question remains unanswered. Will life-forms in other solar systems be constructed on the same chemical basis as that of terrestrial life, that is, principally of carbon, hydrogen, oxygen and nitrogen?

Among the theoretical alternatives, a life-form with silicon substituted for carbon is the most often discussed. Silicon is more commonly encountered on

Earth as silicon oxide, better known as quartz. In the Cosmos, however, silicon is far down the list of the commonest elements: it is among substances which contribute far less than 1 per cent of the total matter of the Universe. There is also another problem. Biochemical reactions must occur frequently enough to produce complex forms by an enormous number of interactions. This is how, on Earth, the replicating macromolecules arise in aqueous solution. (In ice, for example, chemical processes would operate so slowly that evolution would never get very far for lack of time.) The difficulty with silicon is to find a solvent for it that is analogous to water. (In water, it would immediately oxidize to quartz.) Furthermore, no particularly large molecules can be formed with silicon. The exobiologist and astronomer Carl Sagan therefore sees no likelihood of silicon life: 'We cannot make stable silicon molecules that would be complex enough to store anything like as much information as the chain molecules of the genetic code.'

But in crystal form silicon can help to build relatively complex structures. A. G. Cairns-Smith wrote a whole book on the subject in 1971, called *The Life Puzzle*.[41] Indeed, silicate crystals (and crystals in general) fulfil the three necessary criteria with which Oparin would characterize 'life'. In the case of a saturated solution, crystals draw material from their surroundings and use it to grow (metabolism); new (daughter) crystals constantly appear (reproduction) and even generate mutants (mutation), for the new, fresh crystals flourish and grow with other new microscopic rearrangements, which at times produce macroscopically different effects too. But that is about all. Cairns-Smith argues that silicate crystals can store information and can to some extent reproduce their lattice structure exactly. He then suggests an alternative prebiotic evolution on Earth. It began, according to his theory, with silicate crystals, which after a time became capable of utilizing amino acids and finally the nucleic acid RNA. Later, RNA led to DNA in the complex silicate environment, whereupon organisms built on DNA took over from crystals. From then on evolution followed the path with which we are familiar up to the present day. In this hypothesis only the crystal phase of prebiotic evolution is new; it seems neither totally convincing nor obviously wrong.

As a further alternative, a life-form has been suggested in which carbon is combined with chlorine rather than hydrogen—though such molecules have not yet been discovered in interstellar gas. Although the resulting chemical compounds are somewhat exotic, a system of this kind seems workable in theory. A life-form based on carbon, where water as a solvent is replaced by liquid ammonia, is conceivable for the colder planets like the globes of condensed gas represented by Jupiter and Saturn. Apparently it is also possible to form molecules vital to life at $-180\,^{\circ}C$ in this liquid.

What are the prospects for life based on nitrogen or ammonia, but at normal temperatures? Nitrogen has a chemical disadvantage: it can form only three bonds to unite with other atoms, and is thus inferior to carbon, which has four bonds. Nitrogen cannot, therefore, form molecules capable of storing information.

Another reaction cycle that yields energy is possible if carbon dioxide and hydrogen are inhaled, and methane is exhaled. We need not look too far, merely back into the distant past, to conclude that this type of life was probably once dominant on Earth. The primitive atmosphere, rich in carbon dioxide, methane and ammonia, but practically without oxygen, was ideally suited to those bacteria whose immediate successors we identify in the well-known methane bacteria (see 'The third path of evolution', next page). In order to complete the breathing cycle, the carbon dioxide which has been used up must be regenerated. This could take place by oxidation of methane, a reaction that was later spurred on by solar radiation and photosynthesis. Fossil finds seem to show that unicellular oxygen-breathing organisms, the blue-green algae, must have coexisted with methane bacteria. Perhaps the primitive atmosphere contained ecological niches of oxygen, just as, conversely, there are still niches for methane bacteria on Earth.

Other varieties of life are conceivable, and the world of science fiction is full of them. For the present, the bare four dozen or so interstellar molecules provide strong evidence *against* these theoretical alternatives. The organic molecules in interstellar space are exclusively formed on carbon, the element most essential to terrestrial life. The other cosmically abundant substances also play a part in their composition, and provide for the necessary H_2O: hydrogen (the commonest element in the Universe), nitrogen and oxygen. Carbon is present in nearly all the compounds, for it is very reactive and hence combines easily with other atoms. Astronomers have also noted that the denser condensing interstellar clouds, from which new stars originate, contain a notable proportion of those organic elements which take part in chemical evolution on Earth. The organic finds in meteorites deliver the same message.

And so it is not mere Earthly and Earth-bound narrow-mindedness that has led to the usual assumption that extraterrestrial life—at least, the biochemical sort—must be very like terrestrial life. In all our further discussions on the theme of extraterrestrial intelligence we shall (in this book) start from the premise that life in the Universe owes its existence to the reactivity of carbon and the solvent water.

Exotic life-forms: deep-sea microbes

To leave no doubt: if life arises on other planets, it may well be chemically similar to our own and not appreciably more complex genetically. But its external appearance will almost certainly be different, if not indeed vastly so. The mechanisms of Darwinian selection and subsequent evolution permit only the best-suited organisms to co-exist in a given environment. So long as conditions for hypercyclic self-organization were favourable (and they must have been so for life to develop later), the all-encompassing all-or-none elimination contest of the macromolecules will have selected a genetic code and, in consequence of a historical accident, imposed it on all later life.

Although constrained to the same chemistry and the universally (on Earth) similar biochemical language of the genes, the situation on the Earth demonstrates how radically different types of life can develop in the common biosphere of one planet. An example for this was discovered in April 1977 on the deep-sea bottom to the north of the Galápagos Islands, where a team of American oceanographers under the direction of R. O. Ballard on board the deep-sea submarine Alvin from the Woods Hole Oceanographic Institution were investigating a number of hot volcanic springs, which emerge in this area at a depth of 3000 metres from cracks in the Pacific bottom.[42] To their surprise they found that the immediate surroundings of the springs were teeming with life. There were mussels, crabs, sea snails, sea anemones and oysters 25 centimetres in diameter. This fact is the more surprising in that in the depths of the sea, normally at a temperature of 2 °C, extremely poor life-forms are present. In the mineral-rich water of the springs it is somewhat warmer—about 10 °C. Specimens of the water were taken directly from the point of emergence with special equipment. Analysis of the turbid, milky water showed that, besides silicic acid, hydrogen and radioactive radon, it contained primarily hydrogen sulphide, but only a little oxygen. Then investigators suspected that it was not the heat which attracted the creatures, but indirectly the hydrogen sulphide. It is apparently broken down by sulphur bacteria in the deep-sea springs, without need of the effects of sunlight. The sulphur-absorbing bacteria are at the beginning of a food chain that leads via plankton and shellfish to higher animals.

The exotic nature of life on the deep-sea bottom (see Figure 15) is determined by three physical conditions: there is no light, whereas the water pressure is high, and the temperature on the sea bottom is near freezing point. On other expeditions, using a tubular probe, the Alvin had recovered strangely shaped bacteria, several micrometres long, from as deep as 4400 metres.[43] At the end of 1977 another group of researchers from the University of California caught creatures living in the mud and water of the sea bottom in the lightless environment beneath the ice off the Ross Shelf. These organisms, up to 5 centimetres long, are like little greyish-brown trees, and collect suspended food particles with their branches. They belong to a group of unicellular organisms called foraminifera and have a supporting skeleton of agglutinated sand grains.

The third path of evolution

Hitherto the methane bacteria had been counted by microbiologists as just one type of bacteria among many. These unicellular organisms live principally on carbon dioxide and hydrogen, and convert it into methane. Oxygen, the source of life for the greater part of the living world, is quite useless to methane bacteria,

Figure 15 The extreme conditions of the ocean deeps—no light, high water pressure—drive the evolution of life on to exotic paths. At a depth of 5 kilometres, on the floor of the Atlantic Ocean 350 kilometres west of the African coast, lives this long-stemmed sea pen (*Umbellula*), a type of coelenterate which grows to a height of about 1 metre. When brought to the surface, these animals are destroyed by the difference in pressure. (Courtesy: Archiv Urban.)

and so they thrive only in a few secluded spots: the hot volcanic springs of the Yellowstone Park, the deep waters of the ocean, and also sewage filtering plants and the stomachs of cattle. An analysis by American scientists associated with Carl Woese at the University of Illinois in Urbana showed in 1977 that these bacteria are probably the direct descendants of one of the oldest forms of life, which already existed on Earth thousands of millions of years ago.[44,45]

One reason for linking these methane bacteria with those long-gone Earth-dwellers lies in a particular characteristic of the primitive atmosphere. More than 3000 million years ago, when the evolution of life was just getting properly under way, the atmosphere was dominated by carbon dioxide, methane, nitrogen and hydrocarbons: there was practically no oxygen (see 'The Earth's atmosphere: the minimum conditions for life', page 27). The first life-forms thus cannot have depended on oxygen. Why then did no life arise on Venus, whose atmosphere also contains carbon dioxide? The biophysicist Woese suspects that this can be traced, among other things, to the *consumption* of carbon dioxide by methane bacteria. On Venus, carbon dioxide *accumulated*—as it also did on Earth at first—but then, as a result of the greenhouse effect, it brought about a constant over-heating of the Venusian atmosphere. But on Earth the methane bacteria broke down the carbon dioxide in the atmosphere. This increased the transparency of the atmosphere to heat, the greenhouse effect lost its effectiveness, and at the same time the early bacteria supplied raw material for further organic evolution.

It is certainly probable that methane bacteria and related species dominated in the Earth's early atmosphere. However, this does not rule out the possibility that only the last 2000 million years have seen these microorganisms branch off from other evolutionary paths into suitable environments such as—despite an oxygen-rich atmosphere—later existed and still exist today. Woese and his associates did not find the evidence for great genetic age until they examined the totally different genetic structure of unicellular organisms. They concluded from the profound differences they found that the methane-producing bacteria must be classified in a totally new way—that they must be regarded as a separate branch of evolution. By translating the molecular variation difference into the time necessary for the development of these properties, the investigators were able to estimate that the direct ancestors of the methane bacteria must have arisen over 3500 million years ago as a third life-form. Only two 'paths' had previously been known to be followed by the evolution of life on Earth: they led on one hand to the origin of bacteria and on the other to plants and animals of greater complexity.

The proof of this assertion that there was a 'third form' rests on a molecule of ribonucleic acid called ribosomal RNA (rRNA). rRNA is considered a very ancient element of life, which still contains portions of the primitive replication system. Moreover, rRNA is not directly involved in the metabolic processes of the methane bacteria. This was what made the molecule so suitable for making comparisons with other organisms that do not produce methane. Woese has worked for years on rRNA, whose composition he first established in blue-green algae and a large number of other bacteria. The Illinois team found in their investigations that many components of the molecule—the nucleotides—were identical for all types of bacteria studied as well as for blue-green algae. So the question was, would these similarities also be found in the methane bacteria?

The rRNA of 10 different kinds of methane bacteria was therefore investigated with an apparatus known as a sequence analyser, its genetic components arranged according to their basic patterns, and the results compared with the nucleotide sequences of the bacteria and blue-green algae studied earlier. The result was surprising. The bacteria which produce no methane are largely *similar* to the blue-green algae, but are suspiciously *different* from the methane bacteria. These new family relationships do not correspond at all to the classification scheme hitherto used for bacteria, which follows their external form—they may be long rods, oval or spherical bodies. From Woese's analysis of the genes, it would appear that the outward form is a far weaker criterion for the evolution of species than is the case for higher, multicellular forms of life.

Three properties above all—none of them related to methane production in the cell—render the methane bacteria biochemical aliens. First, no methane bacterium contains the enzyme *cytochrome*, common to all other microorganisms. Then it lacks a component of the cell wall almost all other bacteria possess: *peptidoglycan*. And finally another component of the cell, transfer RNA, which, in other bacteria, has almost the same composition as in mammals, lacks a typical section—specifically in the methane producers.

The evidence that the methane bacteria have taken a separate route in the history of life is not easily dismissed. But essentially all this means is that this branch split off very early from the rest of the evolutionary tree, then doubtless including all the mutations and conceivable intermediate stages towards all the other co-existing life-forms. But they have nonetheless sprung from *one* primitive form common to all terrestrial life. The slow changes of the atmosphere drove a few species of methane bacteria with survival capacity into ecological niches. For every life-form that exists today, millions of other species have perished through selection. In the construction of their genes and the bio-chemistry of their cells, the modern geneticist of today is picking up the trail of their history, rather as an archaeologist does in his excavations. As Woese points out, his discovery justifies the hope that some day these methods will give us a substantially more accurate picture of the historical course of evolution, its diversity and its temporal stages.

If the methane bacteria and their relatives were the dominant life-form in the Earth's 'reducing' primitive atmosphere, why did they not develop further into more complex organisms? More complex life-forms only arose after the reducing atmosphere had slowly changed into an oxygen-rich oxidizing atmosphere. 'The reason,' writes the Boston University researcher Michael D. Papagiannis, 'probably is that the higher energy needs of cellular specialization of advanced forms of life cannot be met without the efficient burning of food products with free oxygen.'[46]

No extraterrestrial life in the Solar System

Besides the Sun, the Solar System (see Table 5) consists primarily of nine planets, their moons, and a large number of asteroids, meteoroids and comets. Almost all these bodies, even the Sun, have been suspected in the course of history of being inhabited by alien creatures. And what has not yet been, may perhaps still come to pass. Some scientists who enjoy looking into the future already envisage colonies on the asteroids between Mars and Jupiter and even think settlement

Table 5 The Sun and its planets

	Mean distance from Sun (Earth = 1)	Orbital period (years)	Radius (Earth = 1)	Period of rotation	Mass (Earth = 1)	Mean temperature (°C) S, surface C, clouds	Composition of atmosphere	Composition of planet
Sun	—	—	109.1	25.38 days	332 000	6000	Hydrogen Helium	Hydrogen Helium
Mercury	0.39	0.24	0.382	59 days	0.055	350 (S) day −170 (S) night	None	Silicon dioxide Iron
Venus	0.72	0.62	0.949	−243 days (retrograde)	0.815	−33 (C) 480 (S)	Carbon dioxide	Silicon dioxide Iron
Earth	1	1	1	23 hours 56 minutes 4 seconds	1	22 (S)	Nitrogen Oxygen	Silicon dioxide Iron
(Moon)	1	1	0.273	27 days 19 minutes	0.0123	22 (S)	None	Silicon dioxide Iron
Mars	1.52	1.88	0.532	24 hours 37 minutes 23 seconds	0.108	−23 (S)	Carbon dioxide Nitrogen (molecular)	Silicon dioxide Iron
Jupiter	5.2	11.86	10.97	9 hours 50 minutes 30 seconds	317.9	−150 (C)	Hydrogen Helium	Hydrogen (molecular) Helium
Saturn	9.54	29.46	9.03	10 hours 14 minutes	95.2	−180 (C)	Hydrogen Helium Methane	Water (ice) Methane Ammonia
Uranus	19.18	84.01	3.72	−11 hours (retrograde)	14.6	−210 (C)	Hydrogen Helium Methane	Water (ice) Methane Ammonia
Neptune	30.07	164.8	3.5	16 hours	17.2	−220 (C)	Hydrogen Helium Methane	Water (ice) Methane Ammonia
Pluto	39.7	249.2	0.4	6 days 9 hours	0.1 (?)	−230 (?)	None observed	Water (ice) Methane (?) Ammonia

possible on comets. Some propose that we should first take a closer look at the asteroids. Perhaps secret observing stations are concealed among them, steered to the Solar System by extraterrestrial civilizations.

What opportunities did life have in the Solar System? Organic molecules were formed in the course of chemical evolution in every part of the nebula which was to become the Solar System; but biological evolution and the origin of living things made greater demands on the environment. Within the Solar System, life of the human type could only have taken root on Earth. But was it perhaps possible that evolution had set out on the grand design of life elsewhere, and then come to a stop somewhere halfway, perhaps at the level of unicellular organisms or simple plants?

In this respect the Moon, our nearest heavenly body, proved most unrewarding. The Moon has too small a mass to be able to retain an atmosphere suitable for life. For a long time one of the mysteries of the Moon was its far side, which it kept hidden from our eyes. Speculation had gone so far as to suppose that there might be landing places there for extraterrestrial civilizations. Then a Soviet spacecraft circumnavigated the Moon and photographed its far side. Finally, in 1969 (on 20 July), the first man stepped on to the Moon, and since then the Apollo mission has brought nearly 8 hundredweight—about 400 kilograms —of lunar rock back to the Earth. (The Russians, with their unmanned Moon expeditions Luna 16 and Luna 20, brought back altogether 150 grams of moondust.) The question we are now concerned with is, how was the Moon formed, and were any traces of life found there?

Numerous theories about the Moon have been advocated, and some are still under discussion. Was it thrown out of the Earth (the Pacific is about the size of the Moon) or did the Earth capture the Moon when it happened to pass close by, or did it originate from the proto-planetary gas disc together with the Earth in the manner of binary stars? Selenologists had always believed that analysing a handful of lunar soil would finally solve the riddle of the Moon's origin. But even after exhaustive investigation of the samples from the Moon some problems were still unsolved.[47,48] The fission theory—that the Moon was torn from the Earth— is today the least hopeful. Not only does it meet difficulties in accounting for the Moon's orbit (and its inclination), but additionally the lunar rocks are too different chemically from terrestrial rocks and probably somewhat older. The binary planet theory also has difficulties with the orbit, and in addition the Earth and Moon should not then have such different densities (the Moon, 3.3 grams per cubic centimetre; the Earth, 5.5).

The most convincing theory still seems to be that the primitive Moon first came into being in another part of the Solar System and was later captured by the Earth. At all events, the orbit as well as the composition of the Moon would thus find a natural explanation. Surprisingly, many of the volatile elements were found in very small quantities in the lunar samples. This fact provides a possible hint as to the Moon's place of birth and may solve another problem as well. The law governing the distances of the planets from the Sun allows for a planet within the orbit of Mercury (the Sun's closest satellite). But this place is not occupied by a planet—it is 'missing'. On the other hand, Mercury moves in a very elliptical orbit—an odd circumstance, for normally the planetary orbits are almost circular. It may be that the primitive Moon and Mercury once moved close to the Sun in nearly circular orbits. The interaction between the two soon reached such a degree that the Moon was moved into an elliptical orbit which carried it near the Earth, whereas Mercury, with four times the mass, merely settled into its present elliptical path.

The lunar material has been examined for traces of life with almost fanatical thoroughness, and rightly so. For it was after all the first non-terrestrial material scientists had land their hands on (if we discount meteoritic rocks). Mindful of the scientific *faux pas* which had been committed with meteorites contaminated by terrestrial organisms, special ultra-sterile laboratories were prepared at NASA's Ames Research Center. Over 3000 tests were carried out with the Apollo 11 samples alone. The results were uniformly negative. Not even advanced compounds of carbon and hydrogen were encountered. The minute traces of carbon had probably been accumulated through meteorites and the solar wind. On account of these negative results the quarantine for returning astronauts (which had been proposed by Carl Sagan) was eventually abandoned after the third Moon landing.

But surprisingly enough it turned out that terrestrial bacteria could survive at least for a time on the Moon, although the 120 °C temperature during the lunar day falls to − 180 °C during the lunar night in a 2-week rhythm. In 1969 the Apollo 12 astronauts brought back a camera taken from the Surveyor 3 probe, which had landed automatically on the Moon 2½ years earlier. Scientists in the laboratory found on the camera streptococcus bacteria, which had clearly survived unscathed on the Moon for the whole interval. A Moon puzzle already discovered by the Lunar Orbiters is its shape: the Moon is like an egg with its small end pointing towards the Earth.

If the Moon once orbited near the Sun, its immediate planetary neighbour would be Mercury. In March 1974, the Mariner 10 space probe transmitted back the first close-ups of the smallest planet in the Solar System (if we exclude the Moon). At the first glance Mercury, dotted with meteoritic craters, is so like the Moon that it could be taken for it. Likewise, the surface gravity of Mercury, a third of the Earth's, is too small to retain an atmosphere. And Mercury has been liberally bombarded by meteorites in the course of its history. But the bombardment lasted at most 600 million years, and ended about 4000 million years ago. Mercury shows no sign of erosion by water, which must have risen to the surface during the day if there were subterranean supplies of water. During the planetary day, which lasts 2 months, the temperature on the day side rises to 400 °C, and radiation back into space refrigerates the surface to beyond − 200 °C at night.[49] The surface displays craters and basins up to 1300 kilometres in diameter, and also other ranges and faults. There are indications that the planet must have shrunk considerably as it cooled, and thus acquired 'wrinkles' several hundred kilometres long.

The reason for this shrinkage is the chemical composition of the planet. About 80 per cent of Mercury's mass consists of iron: the iron core probably commences only 640 kilometres below the surface. The high iron content of Mercury is explained by the history of its formation. Before the planets condensed only the lighter elements were carried to the edge of the rotating Solar Nebula; the heavier materials remained near its centre. Currents of liquid metal in the interior of Mercury are probably also the explanation of its magnetic field. But since Mercury rotates only very slowly (during two circuits of the Sun it turns on its axis only three times), its field is very weak—only a hundredth of the strength of that of the Earth. The rotation of Mercury is governed by the tidal effect of the Sun: Mercury rotates in a 2:3 resonance with the Sun. As far as we know, Mercury is just as sterile as the Moon.

Venus, the nearest planet to the Earth and, with 82 per cent of the Earth's mass, closest to it in size, has often been called the Earth's twin sister. Seen from the Earth, it is the brightest of all the planets, appearing as the 'Evening Star' or the 'Morning Star'.

A dense white cloud layer shrouding the entire planet is responsible for the brightness of Venus. But unfortunately the cloud cover also prevents astronomers from getting to the bottom of Venus's secrets. Although many probes (seven Soviet and three American spacecraft) have taken aim at the planet, circled it and in some cases even landed (the first landing was by the Soviet Venera 3 on 1 March 1966), at present there are only two photographs of the surface. They were taken by the Soviet probes Venera 9 and 10, which landed on 22 and 25 October 1975. Venera 9, however, landed on a slope, so that the wide-angle cameras only commanded a haphazard section of the surface. The pictures were fuzzy and bore regular streaks, since the telemetering system of the probes had also to transmit other data. It took several months of work before Luciano Ronca at Wayne State University in Detroit, in collaboration with the Soviet Venera team, had rendered the pictures usable by means of numerous computer processes, which removed the distortion, filtered out the haze and filled in some of the streaks.[50] The results were astonishing (see Figure 16).

Above all else, visibility on the surface of Venus was much better than expected. According to Ronca: 'The illumination was like that on a gloomy, rainy day on Earth.' Venera 9 found on its slope a number of sharp-edged basaltic rocks, 50 to 70 centimetres in length and 15 to 20 centimetres high. Actually it had been anticipated that wind erosion would long since have ground down any angles. But in the 53 minutes during which the equipment on board Venera 9 remained in operation it had only detected light winds of at most 2 kilometres an hour. Venera 10, which set down 3 days after Venera 9 on an almost level spot 2000 kilometres away, gave similar results. Venera 10 ceased

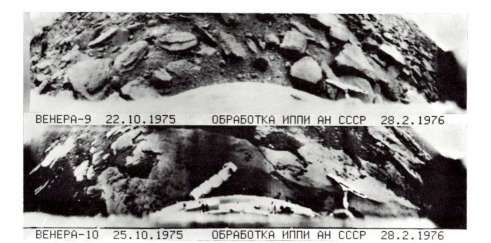

ВЕНЕРА-9 22.10.1975   ОБРАБОТКА ИППИ АН СССР 28.2.1976

ВЕНЕРА-10 25.10.1975   ОБРАБОТКА ИППИ АН СССР 28.2.1976

Figure 16 Immediately after landing on 22 and 25 October 1975, the Soviet probes Venera 9 and 10 obtained the only close-up pictures we have of the surface of Venus. Venera 9 (upper picture) landed on a slope of about 30 degrees and recorded a strip of surface stretching from horizon to horizon (visible in the corners) and narrowing directly in front of the camera to a width of 1 metre. The foot of the capsule is visible in both photographs. The fragments of rubble, up to 0.5 metre in size, are probably lava. At both landing sites, about 2000 kilometres apart, the temperature was 460 °C and the atmospheric pressure was 90 times that on Earth. (Source: Soviet Academy of Sciences, *Bulletin of the Geological Society of America* **88** (1977), 1537.)

functioning after 65 minutes. This was in part due to extreme climatic conditions: Venera 9 registered a temperature of 455 °C at its landing site and Venera 10 measured 464 °C and 90 times the atmospheric pressure at the Earth's surface. Even during the approach by means of parachutes the probes were examining the upper layers of Venus's atmosphere; they found that the Venus atmosphere consists of 97 per cent carbon dioxide enriched by traces of oxygen and water. Together, these gases store up all the solar radiation that penetrates the clouds, and their greenhouse effect keeps the Hell of Venus like an oven. Stormy conditions are met in the upper layers of the Venus atmosphere: yellowish clouds race across Venus at an altitude of about 60 kilometres with a speed of 200 kilometres an hour. They probably consist principally of sulphuric acid—there are occasional rains of sulphuric acid on the basaltic and granitic surface of Venus. Every 4 days the clouds complete one lap of their race round Venus from east to west; the planet rotates quite independently of them within its turbulent atmosphere. Venera 7 and 8, during their parachute descents to the surface of Venus in 1970 and 1972, had already sent the news to Earth that Venus itself rotates much more slowly, and in a most unusual manner, namely 'retrogradely': it turns from east to west or from left to right, the opposite way to the rotation of all the other planets. Not only does Venus rotate backwards, it also turns more slowly than the other bodies of the Solar System. One day on Venus lasts for 243 Earth-days, longer than the 225 days of Venus's year, the time it takes her to make a circuit of the Sun.

The rotational period of Venus, like that of Mercury, is an interplanetary resonance phenomenon, caused this time by the influence of the Earth. As the American astrophysicist Tommy Gold says: 'Venus seems to have started out its life with a retrograde rotation, then it must have been braked by the powerful tidal effect of the Sun, and then it seems to have been caught in a resonance with the Earth. The weak tidal field of the Earth at the distance of Venus was probably enough to hold it in such a way that it always presents the same face at its nearest approach to the Earth.'

Radioastronomers have explored the surface of Venus with radio waves, and found large crater-like basins of volcanic or meteoritic origin. There are, however, no very high mountains on Venus. The high surface temperatures must continue in the interior. But the yielding subsoil cannot sustain high, steep mountains, which would subside relatively quickly. If it had no cloud cover, Venus, like Mercury, would probably resemble the Moon.

The surveying of Venus's surface can be divided into eras before and after December 1978, the month when the American probe Pioneer Venus 1 swung into orbit around the planet. Until then, radio observers had been able to make out only relatively coarse features: a few large craters and two roughly delineated regions in the northern and southern hemispheres, which have been given the names 'Alpha' and 'Beta'. Measurements with Puerto Rico's Arecibo telescope have shown that the Alpha region is covered with parallel ridges similar to the Appalachians of North America. The Beta region, on the other hand, contains a large volcanic crater. Here, too, a canyon several hundred kilometres long was discovered, which strongly resembles the giant rift valley, 'Valles Marineris', on Mars.

It was Pioneer Venus 1 which brought about the turning-point in this laborious process of radio mapping. Since 1978 the probe has explored a full 93 per cent of the Venusian surface: it has found two-thirds of the landscape to be flat, with differences in relative height of no more than 100 metres. Measured against this plain, which serves as a reference level in the absence of an ocean surface, only 16 per cent of the surface of Venus is lower-lying—a small proportion compared with the Earth, whose ocean basins cover some two-thirds of the surface.

The lowest point in Venus's lowlands, which are filled with basalt and probably young, is reached at a depth of 2900 metres below zero in a rift valley at the eastern edge of a majestic upland area, which has been given the name Aphrodite Terra: a range of the size of North Africa. In the other highland mass, Ishtar Terra, Pioneer Venus 1 discovered the highest mountains on Venus, now known as the Maxwell Mountains—a range rising up to 9000 metres above its surroundings and 11 800 metres above zero.

There is evidently only a single continental plate on Venus. The planet shows no evidence of plate tectonics and continental drift. These and related differences between the surfaces of Venus and Earth are now explained by Venus's possessing a thicker planetary crust.

Meantime, the surface of Venus is being copiously supplied with names. Following a decision of the Working Group on Planetary Nomenclature of the International Astronomical Union, larger features are to be named after mythical goddesses, as is already the case for the highlands. Minor features can even be given the names of famous, but no longer living, women.

Why are the atmospheres of the Earth and Venus so different? Calculations show that the carbon dioxide was the same on both planets. The difference is that on Earth the carbon dioxide has been removed by surface reactions, which have incorporated carbon dioxide in calcareous strata. Clearly, the influence of processes which could have transformed the carbon dioxide-rich atmosphere was absent from Venus.

Mars on the biological test-bench

In 1977 Mars was the focus of planetary investigation. On 20 July 1976 the robot probe Viking 1 achieved a soft landing on the surface of Mars, 340 million kilometres from the Earth. On 3 September, Viking 2 also made a touchdown. But these were by no means the first attempts to conquer Mars scientifically. The Soviet probe Mars 3 had landed on Mars on 2 December 1971, an experiment that was repeated on 12 March 1974 with Mars 6. However, both these probes abandoned their task even before landing, and sent back no scientific information to the Earth. But the study by spacecraft of Mars—a planet of half the Earth's size and with about 11 per cent of its mass—had begun much earlier, with the historic fly-by of the American probe Mariner 4 on 14 July 1965. Mariner 6 and 7 followed in July and August 1969, and Mars was circumnavigated for the first time in November 1971 by Mariner 9.

The close-up pictures which rewarded the probes on their missions (see Figures 17, 18 and 19) showed quite enough detail to lay to rest the old myth of the Martian canals, which, among other astronomers, Italy's Giovanni Schiaparelli had brought into the world in 1877. Deceived by fine details in the photographs obtained with telescopes (but which were beyond the resolving power of their equipment) they had thought to discern lines dozens of kilometres wide connecting up sizeable dark spots. The possibility of an artificial origin of the canals suggested itself at once. Was it an enormous irrigation system that covered the whole of the desert planet? The Mariner probes found nothing of the sort. At

Figure 17 This view of the entire northern hemisphere of Mars was assembled from 1500 separate computer-corrected photographs taken by Mariner 9 during its circuits of Mars in 1971 and 1972. The ice cap at the north pole probably extends below the surface of Mars through the whole planet. (Courtesy: NASA.)

Figure 18 The American astronomer Asaph Hall called the two moons of Mars Phobos and Deimos—Fear and Terror— when he discovered them in 1877. His choice of names was taken from the horses Phobos and Deimos, which drew the chariot of Mars, the god of war. Viking Orbiter 1 photographed Phobos, about 22 kilometres in size, on 25 July 1976.[52] Its craters are about 100 metres across. Both moons are probably captured meteorites, which originally came from the asteroid belt between Mars and Jupiter. (Courtesy: NASA.)

best dunes, fissures or mountain ridges might have produced an impression of canals. Instead we saw craters up to 100 kilometres in diameter. These had not been seen with the telescope, because from the Earth Mars is best seen almost always as we see the full moon, under vertical illumination, so that the crater walls cast hardly any shadows.

But other data on Mars, too, were unmasked as errors. For example, the atmospheric pressure had been estimated almost 10 times too high. Instead of the previously supposed 65 millibar pressure (value for the Earth about 1000 millibars) the Viking probes registered only about 8 millibars, eight-thousandths of the Earth's atmospheric pressure. Also, the atmosphere of Mars is mainly (95 per cent) carbon dioxide and not nitrogen, as was still assumed 20 years ago. Nitrogen (2.7 per cent), the noble gas argon (1.7) and traces of free oxygen (0.15 per cent) and water vapour make up the rest—by terrestrial standards an extremely dry and unbreathable atmosphere, through which the Sun's ultraviolet radiation can penetrate without check. Frequent dust storms race across the surface of Mars for days at a time at speeds of 300 kilometres an hour, concealing the deep-frozen desert of reddish (iron oxide-rich) dust and rocks. On the warmest days of the Martian summer the temperature just reaches freezing point (0 °C)—the average for summer is − 23 °C—and in the Martian winter the Viking probes must endure a temperature down to − 123 °C.

This climate is just too warm for carbon dioxide to freeze. In the pressure of the thin Martian atmosphere carbon dioxide would not freeze until the temperature fell to below − 127 °C. Water freezes on Mars at − 50 °C in favourable spots. When measurements of the temperature on Mars were available, the Viking expedition sprang another surprise. The perennial ice caps at the Martian poles are not frozen carbon dioxide, as has always been assumed, but ice: this was established by infrared observations from the Viking orbiters 1 and 2 in orbit above Mars.[51] Mars thus presents itself to us today as a planet rich in water which, bound in permafrost, surrounds a rocky core with an icy shell. Attempts were also made to establish whether and in what way the polar caps grow and melt with the changing seasons. During the 2 months covered by the observations, the edges of the north polar cap shifted at most by a few millimetres. From this it is now assumed that the ice extends below the Martian surface deeper into the planet. Barney Farmer, leader of the team which evaluated the water vapour data, concluded: 'The north polar cap is the tip of an iceberg, floating in a sea of rock . . . .'[51]

Large masses of ice, the remains of water which may once have flooded Mars, would also fit well with other characteristics of the Martian surface. Water on Mars would explain why numerous grooves and channels have been found, which look like dried-up river valleys, some with islands, sometimes even surrounding old craters. In a Martian atmosphere probably 20 times as dense as today, downpours of rain will have washed out the valleys. Water melted by volcanoes may also have flowed from the frozen earth and formed rivers. One

Figure 19 Panoramic picture of Mars (300 degrees) from Viking 1. The view extends from the camera to the horizon about 3 kilometres away. The angular apparatus in the left foreground is the housing of the digging arm, which has not yet been extended. The picture was taken in late (Mars) afternoon. (Courtesy: NASA.)

large rill deserves especial mention. This, perhaps the only true example of Schiaparelli's 'canals' and marked on modern maps of Mars as 'Valles Marineris' (or Mariner Valleys), resembles a magnified version of the Grand Canyon in Arizona. But the Martian version covers the distance from New York to San Francisco.

Exobiologists were of course most eager to know the results of the biological experiments that were to be made by Viking's two remote probes. These were confidently expected to clarify beyond doubt whether biological materials and consequently life-forms—microbes, plants or simple animals—existed on the surface of Mars, or at least had once lived there. But there was nothing on the photographs taken by the Viking orbiters to indicate a Martian biology. However, microbes would not have been detectable at the height of over 6000 kilometres, and it was for such life-forms that the three Viking experiments were to search. Each of the experiments was based on a different assumption about Martian biology. Harold Klein of NASA's Ames Research Center, leader of the biological experiments with the Viking probes, compared it with fishing in unfamiliar waters with three different types of bait: 'We are fishing in a lake in which nobody knows if there are any fish or, if there are any fish, exactly what they may like. . . . A negative result may mean that there is a biology, but that we are using the wrong bait to find it.'[53] It was equally clear that life can exist on Mars even though Viking 1 and 2 obtained no evidence of it at their landing sites 7000 kilometres apart. To carry out the investigation, an arm 3 metres long was to shovel up Martian sand and distribute it in teaspoonfuls between the chambers of the three experimental laboratories.

The *gas exchange experiment* was based on the following simple principle. Living organisms alter their environment when they breathe, take up nourishment and reproduce themselves. Gas exchange can thus provide evidence that life processes are taking place. For the purpose of the experiment a small quantity of Martian sand was mixed with a nutrient solution sometimes referred to as 'chicken soup'. Two hours later the gas above the mixture was analysed. The result astonished the biologists. It contained 15 times as much oxygen as had actually been expected. One day later the oxygen abundance had further increased by one-third. Where did all this oxygen come from? Terrestrial plants consume carbon dioxide and give off oxygen. Had Martian microbes also done this? But Klein remained pessimistic: 'It is too fast for a biological reaction.' However, the chemistry is interesting. On the surface of Mars there must be oxygen-rich chemicals, peroxides or superoxides, which liberate oxygen in an instant on contact with water. This clue to the exotic chemistry of the Martian surface was perhaps the most important finding of the Mars expedition.

In February 1978 the scientists at NASA's Ames Research Center repeated the gas exchange experiment on the Earth. For this they used 'artificial Mars dust', rock dust from the Amoy volcanic crater in southern California. This dust has a high iron content, similar to the Martian material. In the experiment gases were produced in similar quantities to those in the experiment on Mars. The result confirmed the suspicion that all the gases—with the exception of oxygen—are adsorbed on the surface of the dust grains. And the oxygen probably comes from the disintegration of peroxides and superoxides sensitive to humidity.

The *pyrolysis experiment* was on the lookout for Martian organisms which assimilate carbon dioxide. These were to be found by checking whether radioactive carbon monoxide and carbon dioxide added to Martian air in the incubation chamber were processed into organic material. The Martian sample was irradiated by a lamp for 5 days in the radioactive gas, so that possible microbes might synthesize the carbon into organic substances—a process analogous to terrestrial photosynthesis. The subsequent automatic analysis by the probe gave a weak but positive signal. One could indeed deduce the presence of small quantities of organic material. But, warned by the surprising results of the gas exchange experiment, the researchers were not so easily convinced. When the NASA team repeated the experiment with the addition of a little water vapour, the signal that had denoted the presence of organic substances did not appear. This time the results were completely negative. The puzzle of this apparently contradictory reaction has still to be solved, but the scientists believe, on the basis of these and other control experiments, that we are again dealing with unusual chemical behaviour by Martian material.

Finally, in the third experiment, the Martian soil sample was sprinkled with a nutrient solution containing substances previously labelled with radioactive carbon. If, after consuming the nutrients, Martian creatures breathed out the carbon they had taken up, the gas above the sample would gradually become radioactive. Since similar reactions can also occur chemically on Earth, a control experiment had to be carried out after the first one, to exclude just this possibility. After the first run, the sample was sterilized for 3 hours at 160 °C, and the whole procedure was then repeated. The heating should have killed off any genuine biological reaction, the biologists thought, and only a chemical reaction could yield a signal at the second run. When the experiment was carried out at the two Viking stations, it first gave a positive reaction, but on the second run, after heating, the radioactive gas did not appear. Was this the proof of life on Mars which had been sought? Had we finally tracked down the Martian creatures we were looking for? Again the scientists were cautious, and attempted to vary the experimental conditions. When the experiment was repeated, they reduced the sterilization temperature to 44 °C. But the result remained the same. Only when the temperature was lowered to 18 °C, still pretty hot for Martian conditions, did the reaction repeat itself on the second run. Even here, in the scientists' opinion, the result is not unequivocal. On account of the strange Martian chemistry found elsewhere, they think it is conceivable, if not probable, that some material—stable at 18 °C but unstable at 44 °C—is responsible for the peculiar, but probably chemical, reaction. But that it has a biological origin is not excluded with absolute certainty. The likelihood seems small, but a final decision will be reached when Viking 3, equipped with a vehicle, lands on Mars in the mid-1980s.

Despite these reservations the prospects of finding living things on Mars have fallen to practically zero. The question Klein asked himself, after the evaluation of the Viking experiments in early 1977, was: 'Why is Mars even more dead than we thought?' Formerly we thought of Mars as being still at the edge of the life-supporting zone of the Sun. However, the studies of Michael H. Hart and other scientists in summer 1977 (see 'The Earth's atmosphere: the minimum conditions for life', page 27) indicate definitely that neither on Venus nor on Mars, but only in a narrow zone about the mean distance of the Earth from the

Sun, was there a realistic chance for biological evolution in the Solar System. Klein's thoughts seemed even then to be diverging from the general optimism of the life theorists when he said: 'The ease of chemical evolution experiments in which methane, ammonia and water, under controlled conditions, yield interesting compounds such as amino acids, may have led us to believe that the continuation of that process into reproducing chemicals was going to be rather easy. . . . The results from Mars may be telling us that the origin of life is a lot more difficult or complicated than we originally thought.'[53]

The gaseous planets of the Solar System

About 550 kilometres beyond Mars circles Jupiter, the giant among the planets. All the other planets taken together would not amount to one half of Jupiter. Jupiter consists of the same gases as the Sun, and planetologists therefore see in Jupiter a star which has fallen short—one which has merely not quite succeeded in getting thermonuclear processes under way in its interior.

Beneath a thin atmosphere of hydrogen, helium and methane, with clouds of ammonia crystals, frozen at $-150°C$, drift darker layers of ammonia and hydrogen vapours, perhaps also hydrogen sulphide, visible in the telescope as many-coloured reddish-brown streaks. The interior of Jupiter consists principally of liquid hydrogen, and in its deeper layers even metallic hydrogen. In the core, with about 10 Earth masses (3 per cent of the total mass) other metals and silicates have probably collected.

In Carl Sagan's opinion the colours of Jupiter are evidence of chemical evolution among the elements carbon, nitrogen, sulphur and oxygen. In 1959, therefore, Carl Sagan and the Californian biochemist Stanley Miller attempted to repeat Miller's original evolutionary experiment in the conditions of Jupiter's atmosphere—with comparable results. Organic molecules were formed, amino acids among them. Prebiotic molecules are probably present also in the deeper layers of Jupiter, to which solar ultraviolet radiation, meteorites and the lightning flashes of Jovian thunderstorms also penetrate. In 1969, Cyril Ponnamperuma and Fritz Woeller also undertook an experiment in chemical evolution with a mixture of ammonia and methane gases. Their experiment was rewarded with a reddish substance comprising a mixture of hydrocyanic acid (HCN) and cyanogen ($C_2 N_2$).

These and other compounds probably provide the colour for the most striking feature of the Jovian surface—Jupiter's Great Red Spot (see Figure 20). It will hold three Earths, and travels round the planet with the 10-hour period of Jupiter's rotation. The mysterious spot, whose colour sometimes turns from red to brown, has given rise to many speculations. According to current opinion, it provides a kind of atmospheric safety valve: at this point there rises a constant stream of gases—at $-115°C$ about 5 degrees cooler than their surroundings—

Figure 20 Photograph of the planet Jupiter, taken by Voyager 1 from a distance of 20 million kilometres on 13 February 1980. The sulphur-yellow moon Io is seen passing in front of the Great Red Spot, with Europa further to the right. (Courtesy: NASA.)

which carries with it more complex organic molecules, so giving the spot its characteristic variable colour. Since Jupiter radiates into space altogether 2.5 times as much energy as it receives from the Sun, it is supposed that it is still slowly contracting. A contraction of only 1 millimetre a year would account for the observed energy difference.

It is considerably warmer in the interior of Jupiter than in its atmosphere. This is in part due to a warming 'smog layer' of tiny dust particles, which holds back the re-radiation of infrared light from Jupiter's surface. There are probably also warmer layers of liquid ammonia and liquid water on Jupiter. If there are life-forms in liquid ammonia, Jupiter is certainly a suitable home for them. The atmospheric pressure—no accurate measurements are available—is perhaps as high as on Venus, and the surface gravity is 2.6 times as strong as on the Earth. This, however, probably impedes the development of more complex biological forms of life. But the existence of the simplest microbes cannot be wholly ruled out, according to the results of the evolutionary experiments. Ponnamperuma and his co-workers have at all events shown that certain terrestrial bacteria could survive at a pressure of 120 atmospheres and at temperatures between $+30$ and $-200\,^{\circ}C$.

In 1973 two NASA scientists, Paul Dean and Kenneth Souza of the Ames Research Center, discovered in a well in California a rare type of bacterium which lives in a strongly alkaline sodium hydroxide solution, 10 times more concentrated than any other form of terrestrial life can endure. In fact no sodium hydroxide is anticipated on Jupiter, but a similar type of organism with an ammonia basis could be capable of living without problems in an analogous alkaline fluid containing ammonium hydroxide. The close-ups taken by the Jupiter probe Pioneer 10 in December 1973 showed details of cloud motions on Jupiter, but these give us no information on the possible presence of life. Pioneer 11 (fly-by 1974) and Voyagers 1 and 2 (fly-by 1979) provided little information which could be used to test these speculations (see Part III: 'The Galactic Club', page 235). Only a direct landing on Jupiter might perhaps leave us the wiser.

The scientific returns from the two Voyager surveys proved unexpected and sensational. Firstly, they confirmed beyond doubt that the Great Red Spot is a 'stationary tornado'. Its red colour is probably brought about by the presence of phosphine (PH_3), which could be spun up from the interior of Jupiter's atmosphere by turbulence. Once there, it could be broken down into red phosphorus by the Sun's ultraviolet light. The phosphine, incidentally, would give Jupiter a strong smell of garlic! Besides this, the infrared instruments on board Voyager 1 discovered alongside water molecules, ammonia and phosphine mainly the hydrocarbons methane (CH_4), ethene (C_2H_4) and ethane (C_2H_6).

The second sensation broke over the close-ups of the four largest moons of Jupiter—Io, Callisto, Europa and Ganymede, known as the Galilean satellites after their discoverer. The innermost, Io, took scientists most by surprise (see Figure 21): on its sulphurous surface of red, yellow and brown were the first active volcanoes seen outside Earth. 'The finest pizza I ever saw,' commented one NASA scientist.

Io, about the size of the Earth's moon, is—of those we know—certainly the most volcanically active body in the Solar System. During the fly-by of Voyager 1, eight volcanoes were active; when Voyager 2 arrived 4 months later one of these had become quiescent. Sulphur, oxygen and sodium are hurled out of these volcanoes to a height of up to 160 kilometres at a velocity of almost 2000 kilometres per hour.

However, the mechanism which drives Io's volcanoes differs markedly from terrestrial vulcanism. On Io the phenomenon is probably caused by the strong tidal effect of Jupiter, which constantly stretches and squeezes the satellite's plastic interior. The energy liberated by the process forces its way out through volcanic vents. Accordingly, volcanoes on Io are much cooler than those on Earth, with a temperature of about $20\,^{\circ}C$—which is nevertheless some $150\,^{\circ}C$ warmer than the $-140\,^{\circ}C$ cold of the rest of Io's surface.

The dust blown out by this vulcanism also settles into orbit around Jupiter;

Figure 21 This view of Jupiter's moon Io was taken on 4 March 1979 from a distance of 862 200 kilometres by Voyager 1. The bright irregular patches appear to be younger deposits masking the surface detail. The high colour is believed to consist of mixtures of salts and sulphur, brought to the surface by volcanic activity and other processes. (Courtesy: NASA.)

and this helped with a further discovery—Jupiter's ring. According to researchers at the Max Planck Institute for Nuclear Physics at Heidelberg, the greater part of the material in the ring comes from Io's volcanoes. The ring surrounding Jupiter has a diameter of 260 000 kilometres, putting it 55 000 kilometres above Jupiter's surface, and clearly consists of thousands of millions of minute particles—altogether in contrast to the rings of Saturn. Jupiter's ring, only 30 kilometres wide, circles inside the orbit of the closest satellite, Amalthea, which orbits 181 000 kilometres from the surface of the planet.

Such a narrow ring naturally has a difficult time: as a result of bombardment by protons from Jupiter's magnetic field, it is quite possible for Jupiter's ring to be broken up within the astronomically short time of a million years. This is unless there is some sort of 'refill mechanism' continuously replenishing the losses due to the bombardment.[54,55]

Voyagers 1 and 2 will go on to give the outer planets a careful scrutiny, starting with the next one after Jupiter. Saturn and its rings were looked at by Voyager 1 in November 1980. Astronomers have long been puzzled as to why, of all the planets, the yellowish Saturn, the second largest planet of the Solar System, possesses rings (see Figure 22). In the meantime the rings of Uranus and

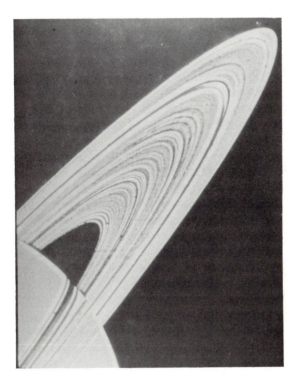

Figure 22 The rings of
Saturn. Contrary to earlier
theories, it is now established
that they consist principally of
fragments of solid material.
(Courtesy: NASA.)

Jupiter have been discovered, so that the mystery of the formation of planetary rings seems to be near resolution (see the next section). Radio observations in 1972 showed that the particles in Saturn's rings range in size from 3 to 40 centimetres. On 12 November 1980, Voyager 1 flew within 124 200 kilometres of Saturn's cloud top and discovered three small new moons, many additional rings of fine material in the Cassini Division (a clearly defined gap between the two outer rings), two very peculiar eccentric rings unlike the usual circular rings, clumps and also spoke-like structures which possibly indicate the presence of electric fields in the rings.

Saturn itself, Jupiter and the other outer planets of the Solar System (Uranus, Neptune and Pluto) consists principally of hydrogen and helium. Clouds course around the planet in brightly coloured bands, and it turns on its axis once in 10 hours and 14 minutes. However, on account of the lower surface temperature of $-180\,°C$ they are less clearly defined than on Jupiter. It is conceivable that the products of chemical evolution, especially compounds of carbon and hydrogen, are suspended in the atmosphere of Saturn.

Titan, largest of the known moons of Saturn, is popularly regarded by astronomers as being in some respects a planet in its own right.[56] Voyager 1 found Titan— a satellite 4820 kilometres in diameter—disappointingly hiding behind a featureless atmosphere, except for a few hazes in the upper atmosphere and a polar 'hood'. Earlier measurements from Earth had indicated a methane-dominated atmosphere of reddish-brown colour, giving rise to speculation about the presence of some form of life on Titan. The sensors of Voyager 1 disproved this theory. Instead Titan's atmosphere is almost entirely made up of molecular nitrogen along with atomic and ionized nitrogen. The claimed methane contributes less than 1 per cent to the total atmosphere and the reddish-brown colour is attributed in part to trace amounts of simple organic compounds such as hydrogen cyanide, produced by photochemical reactions on Titan's surface. Pressure on the moon could be about two or three times that on Earth, at a temperature of about −180 °C. It has been suggested that Titan's atmosphere could be a deep-frozen version of that on Earth at an early stage of the planet's evolution. The highest temperature on Titan—about 30 °C—seems to be reached at the boundary of the atmosphere and exosphere where solar ultraviolet radiation is absorbed by nitrogen.

The relatively hospitable environment on Titan has naturally appealed to writers. In his science fiction story *Imperial Earth*, Arthur C. Clarke made Titan the scene of a large extraterrestrial colony of mankind.

The rings of Uranus

It is only in the last few decades that an accurate prospectus has been drawn up for visits to the more distant planets. Until the eighteenth century we did not even know of their existence: the planetary system came to an end with Saturn. For example, Uranus was first discovered in 1781 by Friedrich Wilhelm Herschel, a military musician from Hanover, then living in England. Uranus moved into the centre of interest again when on 10 March 1977 the surprising discovery was made that not only Saturn but Uranus too is circled by at least five rings (see Figure 23).[57]

Actually the astronomer James Elliot and three of his students from America's Cornell University had set out to observe the planet Uranus itself during a stellar occultation, using a NASA telescope on a transport plane. The planet was to pass in front of the star with the catalogue number SAO 158687. (Occultations provide far more precise measurements than direct observation.) The observations were made during a flight over the Antarctic Ocean south-west of Australia, where the shadow of Uranus cast by this star passed across the Earth for several hours as a consequence of the motions of the Earth and Uranus. But shortly before Uranus itself should have become evident, the light of the star faded unexpectedly several times for some seconds. The same effect was repeated after the planet had passed over the star. Shortly afterwards other astronomers in Australia, southern India and South Africa confirmed the observation. At first

Figure 23 The rings of Uranus were discovered in 1977 by astronomers at Cornell University in the United States. (Source: NASA drawing.)

Elliot suspected that the five long-known moons of Uranus—Miranda, Ariel, Umbriel, Titania and Oberon, counting outwards from the planet—and hitherto unknown smaller objects were the cause of the five slight obscurations. But closer investigation showed that these had occurred with almost exact symmetry of duration and spacing, before and after contact with Uranus, which rendered this explanation most improbable. The different individual moons could scarcely have occulted the starlight so symmetrically by accident; material rings must have been involved. But unlike Saturn's rings they themselves are so faint that in normal observation of Uranus their light is lost, which is why they were not found earlier.

Hitherto the existence of five main rings seems to be established; the rings form a band about 7000 kilometres wide, starting 45 000 kilometres from the centre of the planet (radius of the planet: 26 000 kilometres). The four inner rings are about 10 kilometres wide, the outermost ring about 80 kilometres—considerably narrower than the rings of Saturn, which are up to a thousand times broader. The observational data do not exclude the possibility that there are more rings nearer to the planet. The outer main ring has a peculiar feature. It appears to be 600 kilometres nearer to Uranus on one side than on the other. Either this ring— probably filled with fine dust—is not in the same plane as the other rings, or its form is strongly elliptical.

A by-product of the occultation of a star by Uranus on 10 March 1977 was the discovery of a new moon of Uranus, the sixth. Two Indian astronomers, J. C. Bhattacharyya and K. Kuppuswamy at the Indian Institute of Astrophysics at Bangalore, south India, had observed the whole phenomenon with the 1.02-metre mirror at the Kavalur Observatory.[58] According to their observations a body (still nameless), 70 kilometres across, travels around Uranus a little outside the five rings with a period of 9 hours 20 minutes—the sixth moon. Thus the new moon moves far inside Miranda, the moon hitherto closest to Uranus.

The astrophysicist Tommy Gold—like Elliot—explains the origin of the rings as the result of resonance among the moons of Uranus. If the periods of revolution of the moons have a particular numerical relationship, they can impel small third bodies into specific new orbits. Calculations showed that the rings lie almost precisely in the resonance paths of the two heaviest moons of Uranus, Oberon and Titania. The smaller moon Ariel also moves in such a resonance orbit; perhaps it was formed from a ring that no longer exists, whereas the observed inner rings were not successful in breaking up to form a moon. Gold expresses the suspicion: 'Is it possible that all the moons in the Solar System, just like the planets themselves, originated from rings?' If this is applied to the Earth's moon it would at least explain why it has a different chemical composition from the Earth and therefore cannot have been formed with the Earth. But a ring can have arisen after the formation of the planets, from the material they left behind in the Solar System. Rings of the type circling Uranus or Saturn, however, could not orbit the Earth as we know it today. For that it would have to possess at least one additional moon.

Unexplained perturbations in the orbit of Uranus already pointed to the existence of a still more distant planet in the middle of the last century. The discovery of Neptune is an interesting example of the way in which scientific advances are sometimes made. In 1841 John C. Adams, then a student at Cambridge, set out to prove that a still undiscovered planet was responsible for the peculiar perturbations of Uranus. When he communicated his results to astronomers in 1845 they showed little interest, and no one looked for the unknown heavenly body in the position predicted by Adams. In the same year Urbain Leverrier in France solved the same problem, and predicted a planet in almost the same position as Adams had done shortly before him. But Leverrier failed to persuade astronomers, this time in Paris, to turn away from their stars and galaxies for a while and test his prediction. In his difficulty he finally turned to a young astronomer in Berlin, Johann Gottfried Galle, who finally discovered the planet Neptune on 23 September 1846, less than a degree from the predicted position.

All the giant gaseous planets, up to and including Neptune, may offer opportunities for microorganisms to live in liquid ammonia. But this possibility must remain an unsupported speculation for a long time yet. For the present, a more realistic standpoint may be that we can scarcely expect to find more than a few organic molecules on the planets. This is also true for the final planetary station at the edge of the Solar System, Pluto. When it reaches Pluto's orbit, the light of the Sun has already been travelling on average for 5 hours and 30 minutes. Pluto was discovered in February 1930 by Clyde W. Tombaugh at the Lowell Observatory in Arizona. Here, too, calculations had been tried similar to those which had led to the discovery of Neptune. Despite the systematic search which followed, however, the eventual discovery of Pluto was accidental. In fact, Pluto, 3000 kilometres in diameter and thus less than half as big again as our Moon, is now known to have too low a mass to be responsible for the irregularities observed in Neptune's orbit, and is considered by many astronomers to be an escaped satellite of Neptune. This would at least explain Pluto's elliptical orbit, which deviates so far from a circle that it plainly crosses that of Neptune.

Outside the planetary system, at a distance from 1 to a maximum of 2.5 light-years, move the most distant and least massive bodies of the Solar System, the thousands of millions of comets, made of ice and dust. In the so-called Oort cloud, they move about the Sun in a gigantic spherical shell, just retained by the gravitational force of the Sun. Every year some of the comets leave this cloud and approach the Sun in a flight that takes them many thousands of years. Perhaps some day the comets will provide a target for terrestrial colonists. But as long as we are engaged in a search for extraterrestrial life, we must look until then to the nearer stars.

Habitable planets under other suns

Astronomers face an extremely difficult problem when they set out to track down the planets of other stars. Planets themselves do not shine, and the light of their home stars reflected from their surface is very faint. At a great distance it is drowned out by the far more powerful light of the star itself. Stephen H. Dole estimates[23] that Jupiter could only be observed directly up to a distance of about 1 light-year with the 200-inch telescope on Mount Palomar, even in the absence of an atmosphere and the related interference. The Earth would no longer be discernible with the telescope even one-hundredth of a light-year away. At greater distances, we have only the possibility of detecting planets indirectly by changes they bring about in the stars themselves.

The most promising approach is by the gravitational influence that a planet exerts on its star. Because of the revolution of a planet around it, the star also deviates from its straight-line course and pursues a sinuous path in the sky. This method has been successfully applied to stars: Sirius B, the barely visible companion of the 'Dog Star', Sirius A, was first discovered in this way before it was photographed directly (see Part III: 'The Dogon and the mystery of Sirius', page 273). For planets the effect is of course only measurable for the most massive. Only then does the common centre of gravity, about which the bodies (star and planet) move, deviate perceptibly from the centre of the star. But the larger the planet, the smaller is the chance of finding extraterrestrial life upon it. In other words, these observations are unlikely to tell us anything about the existence of any Earth-type planets, which is where our interest really lies. But it is still difficult to discover planets like Jupiter, and demands photographic and astrometric observations of a star for decades.

A prominent astrometrist is Peter van de Kamp of the Sproul Observatory in Pennsylvania. The star to which he has given the most attention is the second nearest neighbour of the Sun after Alpha Centauri, 5.9 light-years away, known as Barnard's Star in the constellation Ophiuchus. Van de Kamp and his associates observed Barnard's Star from 1916 to 1919, and again from 1938 to the present. In 1963 van de Kamp surprised the scientific world with his claim that Barnard's Star must have an invisible companion, a Jupiter-type planet with an orbital period of 24 years. Six years later he refined his analysis and concluded that it must be a planet of 1.7 Jupiter masses with a period of 25 years. Later in the same year he offered the alternative possibility that there are two planets of 1.1 and 0.8 Jupiter masses accompanying Barnard's Star, with periods of 26 and 12 years respectively. This configuration, according to van de Kamp, would also account for the observed path of the star. If the two planets were transferred to the Solar System they would move in about the positions of the asteroid belt and Jupiter. As Barnard's Star is smaller and fainter than the Sun, we can see in this system a smaller version of our own Solar System.

How difficult it is to learn from the tiny deviations of a star's path from a straight line is shown by the fact that in recent years a regular battle has broken out over Barnard's Star and van de Kamp's interpretation. In 1973 the NASA astronomers David C. Black and Graham C. J. Suffolk announced that Barnard's Star must be attended by a third planet with a period of 7 years.[59] And in the same year two American astronomers questioned the reality of the planet-induced deviations entirely. They attributed everything to over-interpretation of measuring errors on the plates, and maintained that Barnard's Star must in fact be devoid of companions. But van de Kamp still stands by his interpretation. According to his calculations, the nearest Sun-like stars, Tau Ceti and Epsilon Eridani, also have planets. Data for Epsilon Eridani over 34 years would point to a planet with six times the mass of Jupiter and a period of 25 years. Tau Ceti and Epsilon Eridani were made famous by the so-called Project Ozma: in 1960 they were the first to be examined for artificial signals with a radio telescope (see Part II: 'Project Ozma and its successors', page 155).

Where no observations are possible, we must fall back on statistical considerations. In his book on habitable planets[23] Dole lists 14 stars within a radius of 22 light-years from the Sun, near which he believes life-bearing planets to be possible on the basis of theoretical considerations. He assigns the best chances to the star nearest to the Sun, the three-body system of Alpha Centauri. The largest of the three stars, Alpha Centauri A, is quite similar to the Sun (it is a star of spectral class G4 with 1.08 solar masses), and it is possible that a life-bearing planet may orbit around it. Components B and C (Proxima Centauri), however, are ruled out as candidates. But we should not misunderstand Dole's statement about Alpha Centauri A to the point of thinking it very likely that there may possibly be a technically advanced civilization living there, which is only waiting for us to discover it.

Nonetheless, Alpha Centauri is a stellar system in which, despite the presence of three suns, a habitable planet *could* exist. This contradicts another long-standing dogma. It had long been assumed that stable planetary orbits are not possible in multiple stellar systems, in particular double stars. Only for double stars with average separation of the components, corresponding to an orbital period of a day, does this still retain validity. However, if the components are either close together or far apart the orbit remains stable, and the intensity of the radiation falling on the planet does not fluctuate too much. Alpha Centauri A and B, for example, are far enough apart (and C moves at a still greater distance from A and B) that the planet in question could move around star A—as the Earth does around the Sun—without being much disturbed by star B. On the other hand, for two stars very close together a stable planetary orbit would lie at some distance from the close double star, relating as if to a single centre of gravity.

In the meantime it seems fairly clear from the results of researches in 1977 that widely-spaced double stars are of especial interest as regards planets. An investigation in 1977 by the astronomers Helmut Abt and Saul Levy at Kitt Peak National Observatory in Arizona showed that of 123 stars like the Sun 88 could be regarded as multiple systems. A further result is that in wide pairs practically *all* of the components are likely to have additional close companions, either a smaller star or a 'black dwarf' (a stellar body which has too small a mass to start thermonuclear fusion, and—like Jupiter—is providing energy by contraction) or a close planetary system. To the German scientific journalist H. M. Hahn this result means that 'every Sun-like star [in a wide binary] has a companion, which may be a star, a black dwarf, or a planetary system—indeed *must* be, if it is to become a stable star', and that 'wide [stellar] pairs are more than likely to possess planets, just as our Sun may well have a distant partner'.[60]

Has the Sun a companion star?

This was the title given by the astrophysicist E. R. Harrison of the University of Massachussetts at Amherst to an article he published in the British journal *Nature* in November 1977.[61] When this speculation appeared—it was also aired in the December 1977 issue of the German monthly *Bild der Wissenschaft*—it drew attention to a possibility which was already being given serious consideration at Amherst.

Harrison's suspicions were aroused by a number of pulsars, in which he was struck by a remarkable thing: they emit pulses at a constant rate, although according to current pulsar theory they should be quite unable to do so. Pulsars—also known as 'neutron stars'—are bodies of solar mass, exceedingly compact and rotating very rapidly. The neutron material of these bodies, which are about 20 kilometres in size, is as densely packed as in atomic nuclei. The most rapidly rotating neutron star known rotates 30 times a second on its axis. With the same periodicity the radiation cone that the star emits along the axis of its dipole magnetic field sweeps over us, as from a lighthouse.

Pulsars represent the most accurate clocks in the Universe. Their periods vary by less than one-hundredth of a thousand-millionth (10^{-11}) of a second. Thus we can determine with equal accuracy that neutron stars are steadily slowing down their rotation, since the pulse energy is drawn from the rotation. Accordingly the 'note' (frequency) of the pulsation rate falls, and there should not be any pulsars whose note only grows 'flatter' very slowly or even remains constant, for this would mean that the pulsars were no longer losing rotational energy and thus could no longer radiate pulses. A pulsar of constant period must therefore actually be 'dumb'.

But Harrison found that five pulsars had exactly this behaviour. For some of these pulsars the period is diminishing too slowly, for one the period seems to be practically constant, and another seems actually to be increasing its rotational velocity—both circumstances for which today no plausible explanation exists. Another fact disturbs Harrison: all five pulsars are almost in the same direction in the sky—a very improbable clustering.

Harrison suggests a solution for the dilemma: the Sun does not move in a straight line, but is actually circling around another star-like body that has not yet been located. For if the Sun, perturbed by this body into a curving path, is being accelerated precisely in the direction of the abnormal pulsars, their periods could appear to be constant (though in reality they are diminishing), or raised to exactly the values that we observe. The necessary acceleration would be very small—one-millionth (10^{-6}) of the acceleration that the Earth experiences in its path about the Sun would suffice. If in fact another body is affecting the Sun as in a binary system, and if it has about 1 solar mass, it should lie about 50 light-days away in space in the direction of the constellation Vulpecula ('the little fox'), between Cygnus and Lyra. The period of revolution of the Sun about the hypothetical star would then be about 10 000 years.

Doubts naturally arise here. Should we not have discovered so near a companion of the Sun long ago? Would not the Sun's planets, too, have been so strongly perturbed in their orbits that we must have observed these irregularities long ago? A 'silent' neutron star or a black hole are ruled out—they would long since have betrayed themselves by X rays which would have been radiated by high-velocity particles of interstellar gas in their gravitational fields. But a dark body like a 'black dwarf' (a cool star in whose interior no nuclear reactions are taking place) could have been overlooked. And perturbations of planetary orbits, as early calculations showed, would be well within the level of accuracy with which we at present know their motions. But astrophysicists have yet to come up with a reasonable theory which fits this model and explains how the black dwarf and the Sun could have come together as a binary in the first place. This problem does not arise if the unknown body is not connected with the Sun as a double star, but is accidentally passing near the edge of the Solar System in the course of its lonely wanderings through the Galaxy. In that case it would only affect the Sun gravitationally for a few tens of thousands of years during the interval of closest approach.

Meanwhile alarmed theorists are looking for ways of refuting Harrison's ideas as powerfully as possible. Perhaps this can be done with the help of the comets, which surround the whole Solar System—thus also the dark stranger—in a cloud at a distance of about 2 light-years. If there is indeed a companion star, the comets should probably move in orbits quite different from those we observe, when they approach the Sun.

These questions are still too new for the situation to have been clarified. Harrison comments: 'I find it hard to believe that a star so close can exist and yet remain undiscovered. On the other hand, pulsar observations of extraordinary precision imply that it might exist, and therefore a search for a companion star is perhaps worth undertaking.'[61]

The number of advanced cultures in the Galaxy

Let us turn to the question of how observations of planets orbiting our stellar neighbours can help with statements about the probability of extraterrestrial life. The considerations that follow are aimed at estimating the number of technologically advanced civilizations in the Galaxy. Three fields of modern science contribute significant results in this regard. The foundations of the calculations are astronomical theories as to how frequently stars come into being in the Galaxy, how many of them possess planetary systems at one time or another, and near what stars a planet is formed in the life-supporting zone. The second step of the argument is provided by stellar chemistry, molecular biology, and the theories of the evolution of life: on which of the favoured planets does life actually appear and perhaps develop to an intelligent civilization? The conclusions of sociology give us the capstone: rules to describe the rise and fall of technological cultures; laws that permit us to estimate how long on the average such a civilization will survive before it perhaps destroys itself.

Frank Drake and Carl Sagan have assembled this chain of probabilities for the various processes in the form of an equation, which is shown in Table 6. At the conference on 'Communication with Extraterrestrial Intelligence' held at Byurakan Astrophysical Observatory, Erevan in Armenia, USSR from 5 to 11 September 1971, the greater part of the time was devoted to analysing in detail

Table 6

The number of advanced technological civilizations in our Galaxy (N):

$$N = R_* \, f_g \, f_p \, n_e \, f_l \, f_i \, f_a \, L$$

where

$R_* =$	the mean birth-rate of stars in the Galaxy;
$f_g =$	the fraction of stars resembling the Sun which are not members of binary or multiple systems;
$f_p =$	the fraction of such stars with planetary systems;
$n_e =$	the number of Earth-like planets in each system falling within the life-supporting zones of their stars;
$f_l =$	the fraction of these planets on which life has arisen;
$f_i =$	the fraction of life-bearing planets on which life has developed intelligence;
$f_a =$	the fraction of intelligent civilizations which have developed an advanced technology; and
$L =$	the average lifetime of civilizations with advanced technology.

---

the individual terms of this equation (see ref. 62). In the end this should give us a trustworthy estimate of the number of highly developed civilizations, by which in turn we govern our search and sending programmes. Of course we know the least about the lifetime of an advanced civilization(L), since we ourselves know only one such civilization (our own), and even there we have no information about its 'end'. And we cannot base theory on a single civilization experiment which has not even come to an end.

The numerical values that are put into the Drake–Sagan formula represent a mixture of half-established knowledge and ignorance as well as speculation. For we have good information about some of the factors, and none at all about others. Best known is the rate of star production: about 10 new stars a year are formed in the Galaxy ($R_* = 10$). Optimistic astronomers start from the principle that at least every second star is accompanied by a planet. Less certain are the estimates of the number of stars with planets for which a planet like the Earth moves in the ecosphere (n_e), so that its surface provides conditions suitable for the origin of life. The more detailed knowledge about individual processes there is, the more can special factors be introduced into the formula for them. Dole, for example, subdivides the factor n_e into further component probabilities: the probabilities that the planet has the right mass, a suitable axial inclination, appropriate rotational period, and so on.[23] So the Drake–Sagan formula is at present only to be understood symbolically. Particular new findings will have to be taken into account as additional factors. The less statistical information is available, the more random, summary and dubious is the value of the probability taken to characterize the actual frequency of a given process. For example, it seems to me that a great abyss separates the origin of life and the evolution of an *intelligent* civilization. The evolution of life can pursue so many different paths that the chances for the development of organisms with a human-type nervous system seem extremely small. Even on Earth—the sole refuge of intelligence so far known—a visitor from the Universe would only have observed the presence of intelligent life-forms within the last 100 000 years at most—a vanishingly small fraction of the 3000 million years during which there has been life on Earth. The optimistic view nevertheless supposes that more than 10 per cent of all inhabited planets should also have developed intelligent life-forms. The same probability is then assigned to the next transition, from intelligent species to technological civilizations capable of interstellar communication. My doubts concerning these figures are described in more detail in Part II. For these reasons and uncertainties I will not discuss the Drake–Sagan equation any further here. At present we may consider it mainly as a summary of our ignorance allowing no realistic estimate for the number of technological civilizations.

Finally let me make a special comment on the quality of all estimates of probability in connection with extraterrestrial life. The less firmly is a pronouncement rooted in the fruits of our present experience, the more will subjective opinion provide a substitute for knowledge. But the results need not on this account be nonsensical, for although the conclusions are coloured by personal opinion, they are not arbitrary. Although it is not possible to criticize a single subjective probability assumption, we can examine a group of such conclusions for self-consistency and thus to some extent criticize. In the opinion of the probability theorist Terence Fine of Cornell University, 'the concept of subjective probability is at present the only basis upon which probability statements can be made about extraterrestrial intelligent life'.[63]

The figures that different authors obtain for the number of technological civilizations in the Galaxy must be regarded with these general limitations. A suspected lifetime of about 10 million years and a figure of 10 million civilizations are more or less representative. The discussion in this section of the book has shown that the hitherto accepted figures for the ecospheres of the stars (n_e), the origin of life (f_l) and the development of intelligence (f_i) must be considerably reduced. The discussion of the individual probabilities in the Drake–Sagan formula will be taken up again in Part II in the section 'Are we the occupants of a zoo?' (see page 214). Besides this, Part II will present the abundance of astronomical observations which render it increasingly improbable that the Galaxy is inhabited by supercivilizations. Linking the arguments from both sides (molecular biology and astrophysics), I consider it 'subjectively probable' that we are the only technological civilization in the Galaxy (see Part II: 'Conclusion: we are (still) alone', page 222). Stronger statements—supporting this or other points of view—can, of course, be made, but I do not consider them scientifically justifiable.

Nonetheless it will be useful in the discussions following in Parts II and III to have at our disposal as a vocabulary a rough classification of possible types of technological civilization. The Soviet astronomer Nikolai S. Kardashev suggested in 1964 that usage of energy could be taken as a measure of the technological level of a culture.[64] Kardashev distinguishes three types of civilization:

Type I: technological level about that on the Earth; energy consumption about 4×10^{12} watts.

Type II: civilizations which tap stellar energy; energy consumption about 4×10^{26} watts.

Type III: civilizations which control energies comparable to the radiation of a whole galaxy—about 4×10^{37} watts.

In Part II we shall start from the assumption that the Universe is inhabited by all three Kardashev types of civilization. Do we stand a chance of coming to an understanding with them? Have we perhaps been discovered by them already? Are they influencing our fate? Or must we prepare ourselves for the outcome that our longing for 'brothers in space', for an almost all-powerful supercivilization, will remain unsatisfied?

part II
Interstellar exchange of information

Methods of making contact

The idea that the inhabitants of Earth are but one of many forms of life existing in the Universe is as old as the conjecture that many of the thousands of millions of heavenly bodies may be habitable. The ancient Greeks thought it possible that beings might live on the Moon; the astronomer Johannes Kepler and the science fiction writer Jules Verne thought the same. And the German astronomer Wilhelm Herschel speculated in the last century on how things might be for inhabitants of the interior of the Sun.

However, since the turn of the century the Earth's satellite has no longer been considered a habitable place in view of its total lack of atmosphere, and interest has turned to the more distant planets, especially Mars and Venus. But the more we have learned about these and the outer planets, the slimmer has grown the hope of ever coming upon more than a few lowly forms of life within the Solar System: perhaps microorganisms with an ammonia-based biology on Jupiter, or maybe primitive vegetation on Mars, somewhere far from the landing sites of the two Viking probes. Thus, the present century has seen the end of all hopes of discovering life-forms elsewhere in the Solar System with which we could set up an exchange of information. Today's expectations are therefore directed all the more strongly towards the (supposed) multitude of intelligent civilizations on other stars. Among these civilizations astronomers are of necessity interested only in those 'intelligent' enough to build radio transmitters and use them to broadcast messages into space.

But how are we to establish contact with them, trade information and exchange visits? To decide this question, we can do worse than start from the most audacious estimates currently circulating among exobiologists. According to Carl Sagan and Frank Drake there are several million civilizations matching the Earth's standard of development. This corresponds to an average distance of 300 light-years between neighbouring civilizations, giving a duration of at least 600 years for one round of a question-and-answer process. For all less optimistic estimates of the frequency of the origin of life, a distance of thousands of light-years must be bridged. For comparison: the Solar System has a diameter of one-thousandth of a light-year; the trip to Alpha Centauri, the nearest star, is already a journey of 4 light-years.

So long as the exchange of information is seen as the primary goal of contact with extraterrestrial life, the sending of messages by means of spaceships—or similar interstellar probes—seems of extremely questionable utility (see Part III). In order to provide someone far away with information about a culture, or simply about one person, it is (for most purposes) unnecessary to transport physically the whole civilization or the man himself. It is easy to calculate which carriers of information give the smallest expenditure of energy and at the same time the greatest speed of transmission per unit of transmitted information. Of course, the carriers of information must also be simple to manufacture; they must be easy to send off as well as to collect; and the interstellar medium—hydrogen clouds and magnetic fields between the stars—should deflect them or absorb them as little as possible. Of all the 'particles' known to us today, photons, the smallest energy units of electromagnetic radiation, are the fastest: they move with what is to our understanding the highest velocity attainable in nature—300 000 kilometres per second in a vacuum. Photons can be easily generated in large numbers by a variety of transmitters, and can be sent in any desired direction. Photons low in energy—that is, of low frequency—penetrate the interstellar medium as radio waves; the Galaxy is practically transparent to them. Moreover, the energy in each photon is smaller than with any other type of 'projectile': a photon of wavelength 21 centimetres (corresponding to a frequency of 1.42 gigahertz) contains one thousand-millionth (10^{-9}) of the energy bound up in an electron moving with half the velocity of light. For an exchange of information with distant parts of the Galaxy, electromagnetic waves of a given frequency are the cheapest and swiftest carriers—and at present the only realistic hope of establishing any form of interstellar contact in our lifetime.

Where should we listen?

The problem of interstellar communication raises at least three serious questions:

1. Is it in fact possible—without having to drop all other activities on Earth because of it—to send signals over a distance of several thousand light-years with the power and size of modern telescopes?
2. Which frequency ranges in the electromagnetic spectrum are especially suited?
3. In which direction should we transmit?

The opposite questions arise: on what wavelengths should we keep watch; in which directions should we listen? Finally, there is the problem of communication itself: how can we exchange meaningful information with a totally alien civilization? More precisely, what should be the content of our message—how can we compress the maximum of information about human society, in as unambiguous a form as possible, into the shortest possible transmission? What degree of alien-ness must we take into account in composing the message, so that it can still be decoded, however different the life-form receiving it? Should we be careful not to reveal certain data and facts, treating them as it were as interstellar military secrets? What sort of signal should we be listening out for in the first place? Can we expect to be able to decode such a signal at all, especially if we have only accidentally 'overhead' it, that is, if the message was not meant for us?

The history of the reception of allegedly extraterrestrial signals goes back to the beginnings of radio. When in 1899 the radio pioneer Nikola Tesla heard certain electrical disturbances in the rustlings of his receiver, he at first thought them transmissions from an alien intelligence. 'The feeling continues to grow on me,' he wrote, 'that I was the first to hear a message of greeting from one planet to another.' It is probable that he was only receiving disturbances from the Earth's magnetic field.

In the early days of the transatlantic telegraph, the instruments of the Marconi Company received Morse signals of a much lower frequency than was used by Marconi's telegraph to transmit messages. It is reasonable to suppose that these were Morse signals which underwent several reflections within the ionosphere before arriving back at Marconi's receivers with a lower frequency. But Marconi took these signals for messages from Mars, with the eventual result that systematic attempts were made to detect 'further' messages from Mars during its approach to the Earth in 1924.

The history of radioastronomy, of course, only began in 1930, when the American communications engineer Karl Jansky constructed a large radio aerial to track down the source of the disturbances suffered by long-range radio transmissions. What he found were radio signals emanating from the centre of the Galaxy. At first the scientists of that time were very sceptical; however, after 2 years there was convincing evidence that the radio signals he had detected were of galactic origin. But it took the Second World War and the development of radar to provide radioastronomy with a real breakthrough. Military listening posts were frequently intercepting extraterrestrial radio signals. The direct consequence of this was the building of the first large radio telescopes after the war— for scientific purposes. As long ago as 1951 the radio emissions of interstellar neutral hydrogen were detected at the wavelength of 21 centimetres.[1]

Since the introduction of radio—that is, in less than 70 years—radio technology has become so refined that the quality of Earthly transmissions is now limited only by natural factors over which man has no control. Today we have almost reached the point at which an improvement in the transmitter and receiver will have hardly any effect on what is received. But this time interval for the maturing of radio technology is only a hundred-millionth (10^{-8}) part of the age of the Galaxy; considered on a galactic time scale, this means that a civilization achieves the perfect mastery of radio technology quite suddenly, almost without any transition phase. Thus, the chance of coming upon another civilization that has just completed this leap to radio technology is in fact negligible.

We may expect that every civilization whose signals we may one day intercept will already have reached a far higher technological level. If we could investigate a series of inhabited planets, we should probably find that in almost every case radio communication was either still quite unknown or had been mastered to the point of technical perfection. Mankind almost certainly belongs to the extremely small minority of societies which are just leaving this transitional phase—

perhaps one of the few respects in which the human race occupies a special position.

A real dialogue could only take place between long-lived and patient civilizations. But even then it could hardly unfold in the manner of a telephone conversation, in which question and answer follow each other, but rather as a steady stream of information in both directions. And because of the long time the signal requires to reach its destination, the dialogue would not take place between individuals, since—at least, on 'our' side—the lifetime of an individual would as a rule be much too short for an exchange of information.

How can we set up a first contact? It seems pointless to attempt an information exchange with a civilization at a lower technological level than our own. As they could not receive any of our signals, we should have to go to the trouble of physically contacting them in some way. Establishing contact with civilizations technically far superior to us can often be no easier. Perhaps they have long ago made a systematic search for intelligent life in the Galaxy, and know precisely where intelligent life has arisen. The American radioastronomer Ronald Bracewell of Stanford University has proposed[2] that these civilizations have long since established permanent communication links among themselves in a sort of 'Galactic Club' (see Part III: 'The Galactic Club', page 235). In search of new 'club members', they also direct a multitude of radio beams from transmitters distributed throughout the Galaxy to stars in whose neighbourhood they consider technologically advanced civilizations likely to develop. Bracewell thinks it conceivable that at this moment continuous one-way radio signals are being directed at the Earth by automatic transmitters.

If one day contact is actually achieved, thinks Bracewell, it will not be the first time this has happened in the Galaxy—it will already have occurred many times before. Moreover, he says, we shall not establish contact with them; they will make contact with us, but we must keep alert in order to catch their signals. This attitude of expectancy, which I have called the 'optimistic standpoint', will be critically examined in this and the following section. I believe—to anticipate—that it is untenable.

Be that as it may—in order to intercept such signals we must select suitable ranges of frequencies in which to begin a systematic listening programme. Although it is easier than it appears at first sight to choose from the more than 70 octaves of the electromagnetic spectrum a frequency range suitable for our purposes (see Figure 24), we are faced with the additional problem of picking out the most promising directions from all three dimensions of space. Finally, time enters as a further factor: every star which seems promising to us must be 'monitored' for as long as possible. For each of these factors—frequency, direction and time—a choice must be made, as the number of suitable radio telescopes is still too limited, the available observing time (which means the research funds which must be invested) too short, to carry the survey through with any degree of completeness.

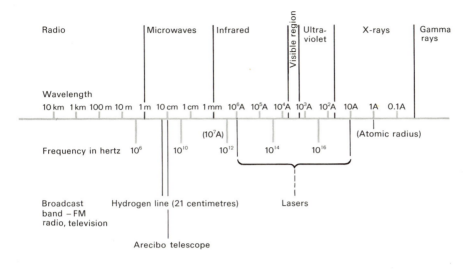

Figure 24 Frequencies of the electromagnetic spectrum.

What is information?

In an article published in 1959, two physicists at Cornell University in America, Giuseppe Cocconi and Philip Morrison, convincingly demonstrated that communication with extraterrestrial civilizations of a level at least comparable to our own is technically possible with electromagnetic waves.[3] Fundamental to this ambitious undertaking are the principles of information theory and communication theory.

Every exchange of the smallest energy units between otherwise separate systems can be regarded as a transfer of information. These systems may be physical or biological; and the exchange, the 'interaction', can take place by electromagnetic radiation, or acoustically by sound waves, or by material particles (Figure 25). Among humans, information can also be conveyed in visible form: in each case the message is converted into a sequence of symbols (an alphabet) understood by the receiver. The telegraph uses the Morse alphabet; man receives electromagnetic waves of certain wavelengths with his eyes; a biological macromolecule like DNA 'understands', that is to say it reacts to characteristic spatial arrangements of movable chains of other molecules by means of their electrostatic force fields. For communication between stars, such as we are considering, it is mainly the properties of electromagnetic waves which are of significance.

In order to reach an addressee, the message must first be converted into a series of electrical symbols, transmitted, received and rid of damage sustained in

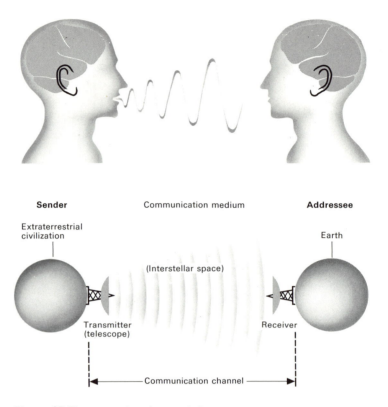

Figure 25 Two examples of transmission systems.

transit, before it can be decoded again into a message for the addressee.

The 'technical' links in this chain, consisting of transmitter (aerial, telephone, etc.), the communicating medium (a cable or interstellar space) and the receiving system, are designated in the abstract by information theorists as the *communication channel*. The Soviet theorist Y. I. Kuznetsov outlines the communication process via the concepts of *communication, coding, signal* and *modulation*: 'A communication [or message] to be transmitted is an entity of information, a concept about some process or about a relationship among phenomena. A signal is a communication converted into a form convenient for transmission. Coding is a method of conversion into a signal. Modulation is a change in the parameters of emission serving as a carrier of the signal.'[4]

The inescapable noise in the communication channel, caused by the instruments and the intervening medium, destroys the original information content with increasing distance. Thus, perhaps the most important parameter in communication with extraterrestrial civilizations is the maximum distance to which a given item of information can reliably be transmitted: the *communication range*. Also significant here is the distance at which the message can still just be picked up, even if its information content can no longer be reconstructed—the *discovery range*.

It is a matter of indifference to the communication channel whether the message to be conveyed is important or trivial—only the amount of data matters. Thus communication theory is not a qualitative concept but a purely quantitative one. The less the receiver is aware what news is in store for him—that is, the less he knows in advance—the greater, as far as he is concerned, is the information content of the message.

Nor does it matter what language, what alphabet, is used. Information of a scientific nature, which principally concerns us here, can be expressed in any alphabet. For the electromagnetic communication of symbols, however, it is simplest to make use of an 'on–off' alphabet, or *binary code*. Electric current offers two states: present (1) and absent (0). The regular flashing of a radio pulsar sends us the binary message: on–off–on–off . . . ; or 101010101 . . . . In pictorial terms, this is a black-and-white pattern, as on a chessboard. Of itself, this message contains practically no information: but it testifies to the existence and position of the transmitter. From the astrophysical point of view this information is nevertheless of value: we can deduce from it the properties of the source of the radiation—possibly a rotating neutron star—and the mechanism which gives rise to the radiation. But what really makes a message interesting in content is information beyond that contained in pulsar signals (Figure 26) or the ticking of a clock, and this must be checked for in each case. Only thus can we be sure to some extent of not being tricked by 'Nature'.

In interstellar information exchange there is no possibility of agreeing on an alphabet beforehand. This agreement must necessarily be contained in the message itself, as far as possible in a self-explanatory and self-evident form. It should be placed at the beginning of the transmission—otherwise the message itself will remain incomprehensible.

Artificial message or natural noise?

When in 1967 the British radioastronomer Antony Hewish and his associates came on the first radio pulsars, the initial suspicion was that some terrestrial radio signal had been accidentally picked up. But when it was clear beyond doubt that the pulses did not come from the Earth but from somewhere in the Galaxy, the excitement was considerable. Had an artificial source, a sort of cosmic beacon, really been detected? The first pulsar was christened LGM-1 (Little Green Man 1).

But astronomers and astrophysicists have become extremely cautious in recent years with respect to the 'artificial' nature of the phenomena they have

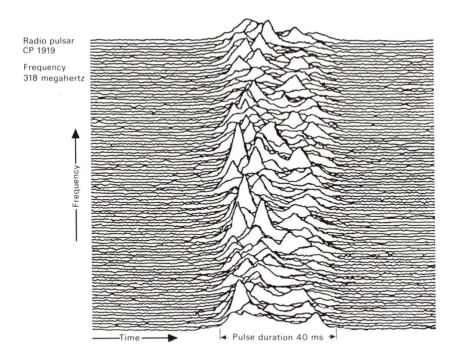

Radio pulsar
CP 1919

Frequency
318 megahertz

Frequency →

←—Time—→ |← Pulse duration 40 ms →|

Figure 26 Radio pulses show a fine structure at different frequencies.

discovered. Only too often have premature statements regarding 'discoveries' of this kind been made, which were later found to have 'natural' explanations— from the historical speculations about the canals of Mars to the UFOs and 'artificial' signals of the 1950s and 60s, which were fed to the mass media with great frequency, only to prove on closer analysis to be 'false alarms'. It has profited astrophysicists, therefore, to use all their ingenuity in search of a natural explanation for cosmic phenomena which were not understood, or to assume that an apparent artificiality had been 'smuggled' in by an unconscious human prejudice during evaluation of the data (see 'The curious waves of gravity', page 179).

The artificial nature of a message declares itself through peculiarities in the properties of the electromagnetic wave packets—that is, their amplitude, frequency, polarization and phase. We can add to this the bandwidth, the amount of the frequency scale taken up by the signal. Many of the known natural sources of radiation, both stars and interstellar gas clouds, commonly display very large bandwidths. But some natural sources emit narrow-band signals, 'characteristic' radio lines, like that of interstellar hydrogen at a wavelength of 21 centimetres. This line signal nonetheless has a bandwidth of around 50 kilohertz.

The narrower frequency band of the hydroxyl (OH) line at 18 centimetres covers only a few hundred hertz. To stand out against these natural radio sources, a civilization still comparable to our own (Kardashev's Type I—see page 90) would probably prefer a narrower band—a few hertz wide or even narrower—for a signal which would be identifiable as artificial. This is also cheaper: signals with greater bandwidth require considerably more energy. Supercivilizations, at best, will have the technological resources to produce artificial signals of high intensity, covering several octaves of frequency, for continuous transmission (see 'Radio contact between galaxies', page 168).

How does a message get into an electromagnetic wave? A message is changed into an electromagnetic signal by *modulating* one (or several) of the properties of the sequence of waves, or wave train—'marking' it in accordance with a definite code, such as the binary code. At the other end of the communication channel, the receiver recovers the information by picking out these marks and retranslating them into a message. To do this, of course, he must know the code. Figure 27 shows the various possible ways of modulating a wave train. Which particular type of modulation we should give most attention to when decoding, or use ourselves in encoding a message for the stars, must be decided by the property which makes its discovery most likely (see 'The search for contact as a two-party game strategy', page 139).

Even if a radio source behaves (statistically) sufficiently 'abnormally', and can be separated from the cosmic radio noise, there remains the big problem of distinguishing it from the mass of natural sources: galaxies, neutral and ionized hydrogen clouds, supernova remnants, individual stars. With a distant astronomical source, whether interstellar gas cloud or extraterrestrial civilization, no code can be agreed upon in advance: it must be worked out by the receiver himself. After he has broken the code—that is, the type of modulation—which the sender has used for the message, the incoming data must be processed by computer in reduction programs. There is little point in applying this procedure to radio sources whose natural character is undisputed. Sebastian von Hoerner of the National Radio Astronomy Observatory at Green Bank, West Virginia has suggested that, with an artificial signal—sent with the intention of being as easy as possible to discover—the message may be preceded by an introductory (or constantly repeated) *call signal*.[5] A call signal would be immediately identifiable as 'artificial' by a particularly conspicuous type of modulation and thus draw attention to itself. It would also at once explain the code in which any following messages would be expressed: an interstellar language tutor, so to speak. This signal should not contain too much information, for its purpose is primarily to reveal the existence and address of the sender and the code used in the message.

It is enough for a call signal to carry the minimum of information which will adequately identify a message as artificial. As a rough guide, it is thought that the call signal should carry perhaps between 10 and 100 bits: no less, so as to preserve

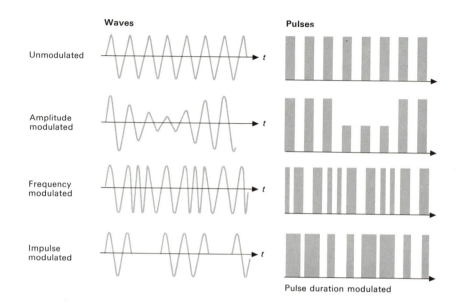

Figure 27 An electromagnetic wave can be supplied with a message by means of 'modulation', the term for systematic changes in the four properties of the wave—amplitude, frequency, phase and polarization. The figure shows a number of possibilities, for both sine waves and pulses. At the top are two unmodulated signals; in the first row below, the amplitude is modulated, here using a binary code: 0 = small amplitude; 1 = large amplitude. This code can also be expressed by changing the frequency: 0 = rapid oscillation; 1 = slow oscillation. A compound modulation expresses the code through breaking up the wave into pulses. For pulses, the pulse duration can also be modulated: 0 = broad pulse; 1 = narrow pulse.

its artificial character; no more, to avoid too much loss of intelligibility resulting from a partial loss of information during transmission. An example of a call signal with around 60 bits would be the first seven prime numbers together with their sum and a character to separate two such sequences on repetition:

$$2 \quad 3 \quad 5 \quad 7 \quad 11 \quad 13 \quad 17 \quad 58 \quad 0$$

In binary code this call signal would be:

000010 000011 000101 000111 001011 001101 010001 111010 000000

In decoding messages it must be remembered that artificial signals may even be *only* call signals and may consequently contain no further information. Frank Drake of Cornell University thinks it conceivable and even possible to 'graft' an artificial spectral line on to the radiation of the Sun, in order to contact all civilizations in the Galaxy.[6] (In a terrestrial context, an 'artificial spectral line' is one which does not occur in the solar spectrum.) This abnormality might attract the attention of astronomers in other civilizations, and could already be achieved with our present technology: several hundred tons of an element with a very rare atomic transition probability would not exceed the carrying power of currently available spacecraft, which could put the material in a close orbit around the Sun. The spacecraft might follow a path similar to that taken by the Helios satellite (a joint venture in which Germany was involved), which approached the Sun far inside the orbit of Mercury. On reaching the point nearest the Sun the material would be ejected in the form of dust, and would distribute itself along its orbit as an elongated cloud. This operation has already been tested in the neighbourhood of the Earth: artificial satellites have often thrown clouds of different materials and metallic particles into high orbits for experimental purposes—for instance, to study their reflecting qualities, or to explore the possibility of interfering with terrestrial radio traffic. The 'lifetime' of the dust cloud will depend on its distance from the Sun. The constant wind of particles from the surface of the Sun and its powerful radiation will heat the cloud and slowly destroy it—the more rapidly, the closer it orbits the Sun. At greater distances this effect would be reduced, but less sunlight would encounter the cloud, making it observable in the solar spectrum only from a few directions.

The solar spectrum is continuous, with over 25 000 absorption lines at present identified. Astronomers assign it to spectral class G (more precisely G2), typical of a star with a surface temperature of 6000 °C. G-class spectra show particularly strong lines for the metals iron and calcium; but there is no sign at all of the noble gases (other than hydrogen and helium), the rare earths, the halogens, or elements of atomic number greater than 79. These substances are probably absent from the Sun, except for small traces, and it is from among these that a suitable element must be chosen. At the wavelengths characteristic for this element new absorption lines, most uncommon for G stars, would appear in the solar spectrum. Although quite weak, they could be detected by extraterrestrial astronomers at a distance of up to several hundred light-years.

The idea of a call signal seems to me a universal one, hardly dependent on special terrestrial patterns of thought. In fact, a considerable part of the total transmission time of radio traffic on Earth is devoted to call signals. They convey no information except the existence and activity of a particular transmitter. And they have the advantage of being economical: the sender need not go fully into action—increase the energy of the signal and narrow the frequency band—until contact has actually been established. The sender would then change from an initial broadband call signal (to be heard by as many as possible) to a strong narrow-band transmission. Once contact has finally been established, a great variety of measures can be undertaken; for now there.is some purpose to it, whereas it is fruitless and frustrating to expend labour and money on transmissions which no one is listening to. Given the general physical boundary conditions to which every civilization in the Universe is subject, it is probable that other civilizations will develop a similar sense of economy. This may even be *demonstrable*, at least for civilizations on our technical level which can afford such an intellectual luxury. Let us assume that the extraterrestrial civilization that is addressing us—accidentally or purposely—has also thought of the possibility of a call signal. Radioastronomers should then be on the lookout for signals of the following kind: narrow-frequency band, quasi-monochromatic, pulses well separated in time and modulated according to a definite code.

The complexity of the problem compels us to make some simplifying assumptions, which indicate clearly defined channels for interstellar communication. But the communication problem is not to be solved by wishful thinking—everything may well be quite different. Information may be transmitted continuously with a large bandwidth. Although this indeed demands a great deal of energy, it also permits the transmission of greater amounts of information. The fact that the signal is artificial can be made evident by special properties of the continuous transmission (see Figures 28 and 30). Apparently with the intention of bringing still more certainty into the search for civilizations, communication theorists have shown that messages can be devised which are deceptively similar to natural signals.

Random numbers, for example. Computers can be programmed to produce series of numbers of any length, fulfilling the following conditions:

1. Each individual number occurs on the average with equal frequency.

2. Each combination of successive numbers (pairs of numbers) occurs on the average with numerically equal frequency. The same is true for all possible combinations of three consecutive numbers (triplets), combinations of four numbers, and so on.

3. Other recurring configurations of consecutive numbers (periodicities) are also uniformly distributed statistically among the sequence of numbers.

If we now extract a small section from a series of random numbers, the frequencies in this (or any other section of finite length) are no longer as uniformly distributed as they are in the longer series. Certain 'structures' will simply appear more often than others—and more conspicuously so, the shorter the selected section is. But just such deviations from statistical uniformity would serve us well as measures of artificiality. When has such an 'artificial' message been fabricated? The Soviet linguist B. V. Sukhotin proposes as a criterion that we should make use of the fact that, in terrestrial languages, combinations of vowels and consonants are more frequent as a rule than in a random series of letters (see 'Pictograms and the computer: automatic image recognition', page 132). In the exchange of messages 'among ourselves' on Earth, this might certainly be helpful.

What is the situation in the search for artificial extraterrestrial radio sources? Natural noise corresponds to the numerical 'noise' of an infinite series of random numbers. Assuming that, for technical or astronomical reasons, we contrive to record only a short section of the radiation of the source, then, on purely statistical grounds, it is not impossible for these data to appear on decoding like an artificial message. But the more artificial these data seem to be, the less probable it is that this is actually the case. For example, we certainly cannot anticipate that the exact second act of Shakespeare's *The Tempest* will be presented by chance in a short stretch of the noise from a cosmic radio source.

This, I hope, illustrates the fact that, in principle, a definition of an 'artificial signal' in the abstract simply cannot be given. Nevertheless, we would orient ourselves instead, more pragmatically, according to specific indications. The criteria for artificiallity can be summarized in two groups:

1. Indications of artificial origin for the cosmic source.
2. Special properties of the radiation received, which have been deliberately stamped upon it by an extraterrestrial civilization, in order to make the discovery and decoding of the signal as easy as possible (existence of a call signal).

Some artificial properties of radiation—modulation and call signals—have been discussed above; but in addition the external appearance of an astronomical object may possibly give indications that a technologically developed civilization exists (see 'The astroengineers of supercivilizations', page 200). Among the most important indicators in the first group is the apparent size of the source in the sky, its angular diameter. Even if a civilization is able to misappropriate its whole planetary system, together with its sun, for the transmission, the angular size of the artificial source will still be relatively small. (This need not be true for civilizations of Kardashev's Type III, which can transform whole galaxies.)

Besides this, the more information is to be transmitted by it in the transmission time, the smaller must an artificial source be. In order that a pulse may be emitted simultaneously from all parts of the transmitting system—for instance, an aerial—the pulses must be fired off at an intervals greater than the time required by light to cross the linear diameter of the aerial. If the interval between pulses is shortened, separate pulses no longer reach the receiver—they would overlap, as Figure 28 shows.

The angular diameter of cosmic objects can be measured with the help of radio interferometry. Here radio waves from the same source are received by different radio telescopes and recorded on magnetic tape. The separate flows of data are then superimposed by a computer and made to interfere. The properties of the superimposition pattern permit the angular diameter of the source to be calculated. The distance between the telescopes is the so-called *baseline*—whence the term *Very Long Baseline Interferometry*. The greater the distance between telescopes, the smaller the detectable objects can be. The individual radio dishes need not be particularly large, but the attainable resolving power greatly exceeds that of the largest individual telescopes. Terrestrial radioastronomers are limited

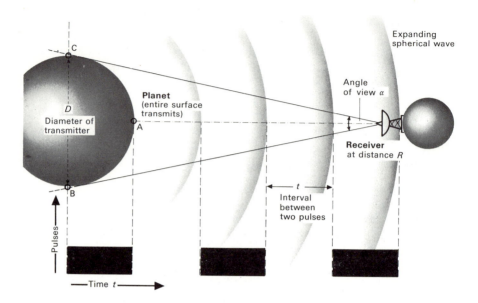

Figure 28 Pulse length increases with size of transmitting aerial: information transfer becomes slower. The radiation from a pulse sent by an extended source (here, a telescope covering half the planet) reaches the receiver first from point A, last from points B and C. (This is also true for a plane telescope.) This reduces the clear interval between adjacent pulses. The receiver sees the source within his

$$\text{Angle of view} = \frac{\text{diameter of source}}{\text{distance}} = \frac{\text{speed of light}}{\text{(distance)} \times \text{(pulses per second)}}$$

This formula makes it clear that the more information is to be transmitted, i.e. the more pulses are to be sent off per second, the smaller the angular size of the source must be.

as regards the distance between telescopes by the size of the planet itself. A tremendous improvement can be expected when a radio telescope is installed in Earth orbit or on the Moon (see Figure 29).

A further indicator of the artificial nature of a radio source is the distribution of the radiated energy across the different frequencies of the frequency band— the radiation spectrum. Natural sources have particular, characteristic spectra. Artificial sources could thus be identified by the differing nature of their spectra (Figure 30).

All the criteria we have mentioned for the artificial nature of a source and its radiation, and for significant information content, can only help us to produce a short-list of possible candidates which will be useful to us; they are necessary but still inadequate criteria. But the universal laws of physics provide the basis for the common language with which the problem of cosmic communication must be solved.

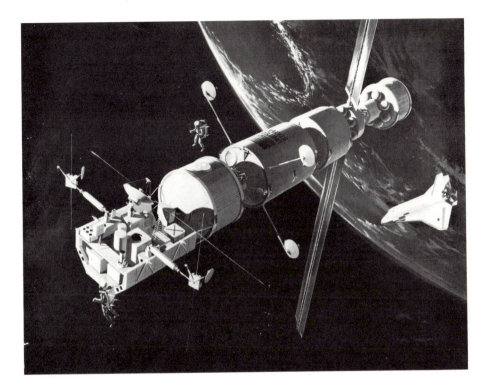

Figure 29 NASA plans to put a large space telescope into Earth orbit in the mid-1980s, using the re-usable Space Shuttle. It will be able to resolve optical sources with an angular size down to one-tenth of a second of arc (60 seconds of arc = 1 minute of arc). This is 10 to 20 times smaller than can normally be made out from the surface of the Earth.

What to send? What to ask?—Message content

When the Arecibo radio telescope was re-opened in November 1974 after a 3-year period of reconstruction, the opportunity was taken to transmit a specially composed message into space (see 'The Arecibo cosmogram of 16 November 1974', page 135). It took less than 3 minutes—to be precise, just 169 seconds—to send the messsage, which was intended to provide basic information about the inhabitants of the planet Earth. After the transmission, a great controversy arose as to why this event had not been announced beforehand and discussed internationally. Some scientists criticized the secrecy with which the plan was implemented, since by a resolution of the 1971 Byurakan Conference (see page 87), such messages should have been drawn up 'by representatives of all

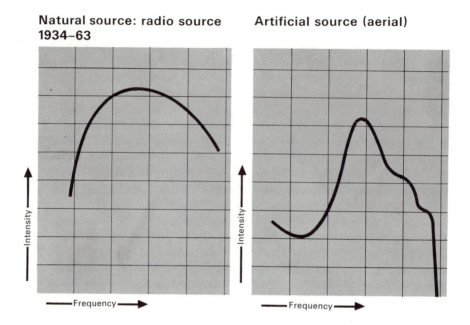

Natural source: radio source 1934–63

Artificial source (aerial)

Figure 30 Spectra of artificial transmitters and astrophysical sources are different.

mankind'. Behind this indignation there clearly lay the desire that mankind should confront extraterrestrial civilizations with solidarity and unity.

At the same time this undoubtedly demonstrated a certain anxiety about the uncertain consequences of contact with 'extraterrestrials', especially if their technology should be superior to our own (see 'Monsters conquer the Earth', page 221). On account of this fear, extreme caution is often urged in the choice of content for the radio signals to be transmitted: should mankind merely call attention to itself by means of call signals, or give away accurate spatial coordinates for 'Sol III', the third planet of a G2 star at the edge of the Galaxy, and risk colonization and extermination? This point may seem extremely Earth-orientated—cosmic contacts seen as magnified versions of the establishment of terrestrial colonial empires. But it is seriously considered by many people, and in several countries censorship keeps a tight control on the discussion of this subject in books for this and related reasons. (As examples, we can mention books about 'extraterrestrials' which have appeared in East Germany and the Soviet Union.)

Nevertheless, most scientists who theorize about the content of interstellar messages feel compelled to devise a sort of cultural passport picture of mankind, and to send it into space. This is intended to give, in the briefest possible form, information about the fundamental character of our biochemistry, the level of our science, the nature of our social, ethical and also religious life—as was partly done in the Arecibo message. How 'wise' is such a procedure? Aside from a certain eagerness to make ourselves known to possible inhabitants of the Galaxy or the immediate threat of annihilation (by a nearby supernova, for instance), I find it difficult to imagine a motive which would lead terrestrial civilization to send out its current information equivalent as a message. Typical time intervals for galactic cultural exchange run to thousands of years at least. Even if we actually succeeded in establishing relations with a group of older civilizations, it is 'a great conceit, the idea of the present Earth establishing radio contact and becoming a member of a galatic federation—something like a bluejay or an armadillo applying to the United Nations for member-nation status'.[7]

This argument of Sagan's does not apply to all terrestrial groupings. The United Nations Organization, for instance, does not seem to impose any minimum cultural or moral standards on potential members. For the next few centuries or beyond, the Earth would be unlikely to derive any direct advantages from its cosmic message sending (see Part III: 'The Galactic Club', page 235).

But if we do decide on galactic transmission of information, we must make a radical start from an intellectual zero point, and from there on, step by step, lesson by lesson, the basic concepts and their relationships, the grammar, must be built up. The beginning, according to an idea of Philip Morrison's, would be made as though between two men who have no common language: they point to objects and name them (see 'Lincos—an interstellar language?', page 120). In radio communication we can 'point' by introducing certain symbols as often as possible in the transmission in different contexts, so that they are learned through their practical application. We can start with mathematics. Using modulated radio waves, Morrison shows how we can begin by introducing arithmetic numbers, representing them by simple pulses. Two such pulses denote the number 2. Differently shaped (modulated) pulses symbolize the signs '+' (addition), '−' (subtraction),'=' (equals), and inversion (forming reciprocals). The corresponding teaching program would take the following form in part:[8]

Addition	$2 + 3 = 5$
	$3 + 5 = 8$
Subtraction	$7 - 4 = 3$
Division	$6 \div 3 = 2$
Reciprocation	$R2 = 1 \div 2$
	$\pi \div 4 = 1 - R3 + R5 - R7 + R9 - R11$

When transformed into pulses this gives the modulated wave-forms shown in Figure 31.

This example is certainly too simple and fragmentary. But it shows the direction in which methodical progress can be made. The very first candidates for message content are the fields of scientific knowledge: logic, mathematics built upon it, physical and chemical laws, the biological basis of our life, the structure

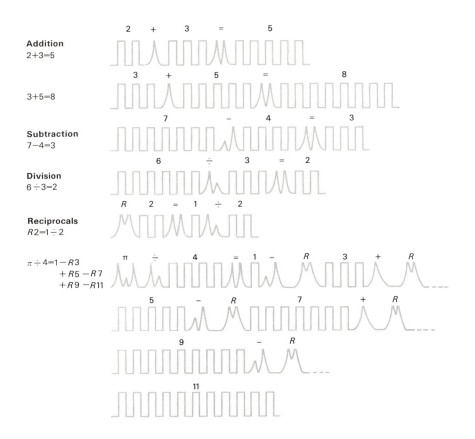

Figure 31 Teaching mathematics with pulsed signals.

of the Solar System, the position of the Earth, the number, appearance and size of its inhabitants. In addition, other civilizations might be interested in the forms of our society, our psychology and philosophy.

Criticisms have at times been levelled at this list of subjects: though, of course, we cannot foresee what might interest other civilizations in us (runs the objection), it will be probably least of all be our scientific knowledge! For if a society is able to receive our radio signals, it must have achieved for itself at least the basic physical and technical knowledge this requires. And if besides (which is probable) it is a substantially older culture than we are, it will not be particularly interested in our (from their point of view) relatively primitive physics. Philip Morrison is himself convinced that a cosmic message 'will contain not mainly science and mathematics but mostly what we would call art and history'.[9] He thinks it easier for a long-lived society to solve scientific problems for itself than to spend time analysing magnetic tapes full of extraterrestrial signals. On the other hand a civilization is totally incapable of reconstructing legendary or historical events.

It is not always easy to distinguish which social and cultural achievements are 'not reconstructable'. This, of course, is due above all to the fact that we have not yet come across any other civilization whose history can be compared with that of mankind to find parallel developments. In particular it is doubtful whether, for instance, mathematics would develop on even approximately similar lines.

The basis of any kind of science is logic. But there is no 'absolute' logic which has universal validity. Mathematics and physics rest on a quite specific logic: the logic of two alternatives, in which the principle of the excluded middle (*tertium non datur*) holds good. That is to say, all statements considered are either 'true' or 'false'—these are the two admissible 'truth-values'; and if I negate something twice, I have affirmed it (a double negation equals an affirmation).

But if we abandon the last property—the principle of the excluded middle—we come at once to another version of the logic of two alternatives: 'intuitionistic' logic. A mathematics can likewise be constructed on this logic, and L. E. J. Brouwer, Hermann Weyl and others have attempted this since 1907. Compared to ordinary mathematics, the resulting intuitionistic mathematics is 'less structured': since we have one less tool (*tertium non datur*) at our command, fewer propositions can be proven.

Besides these two-valued logics, however, there are also logics with more than the two alternatives 'true' and 'false': the so-called 'multiple-valued' logics. And for each multiple-valued logic, as for two-valued logic, there are many possible versions. But for interstellar communication it should suffice that both transmitter and receiver have at least some kind of logic. It is possible that many life-forms get along without any logic at all—but how then can we come to an understanding with them? Logic relates to discussion between individuals: it provides rules according to which conclusions can be drawn. Without logic there is no communication.

In order to make sense of a message from space, we are forced to start from the assumption that it is based on a logic whose rules do not change in the course of the message. Then the statements between two systems of logic can be translated from one system of logic to another.

If we include mathematics in the 'non-reconstructable' part of a culture, the findings of chemistry and physics are still of interest to another civilization, even if it is already in possession of this knowledge: different fields of mathematics may have developed to different extents. If a message indicates an area which is as yet underdeveloped, we have an incentive to pursue it more thoroughly.

I differ with Philip Morrison in thinking that the communication of common knowledge is essential. The lowest common cultural denominator is to be found in the material constraints, relatively independent of cultural factors, involved in the technique of transmitting messages. However obvious this information may be to both sides, it is the testing ground on which an understanding can most easily be achieved. If we began at once by transmitting unique cultural information, such as the history of literature, data for which there is absolutely no common ground of experience between civilizations, we should probably be unable to make ourselves understood at all. Only when based on shared scientific knowledge has the unique and unreproducible content of a civilization any chance of being understood.

But what would *we* wish and be able to ask? What questions would at this moment be the most important for mankind, so that the answers would have the most enduring influence on the future of our society and our science?

Questions about the galactic position, the chemical basis and the technological level of the civilization immediately suggest themselves. But beyond this we should be cautious. Not every interesting question is a good question. The briefer and above all the more precise, that is, more detailed a question is, the easier it is to answer. With imprecise questions we risk receiving an answer like that of the Delphic oracle.

Good questions would be those about physical and mathematical problems which are puzzling scientists on Earth today: the theory of elementary particles; the relationship between quantum mechanics and Einstein's General Theory of Relativity; the general solution of the Einsteinian gravitational equations. Or cosmological problems: Is the Universe open or closed? What was physics like during the first fractions of a second after the Big Bang? What happens to matter in the interior of black holes?

But my physicist colleagues, among whom I took a small private census about the questions they would like to ask, considered none of these questions urgent enough to be the first to be directed to an extraterrestrial society. Those most frequently mentioned were: 'How can the energy problem be solved?' 'How can our civilization survive its present stage of technical development without destroying itself?' 'What is the structure of the society of the future—of a civilization, that is, which may have survived similar destructive phases and remained stable against the crises of civilization?'

Other questions were mainly aimed in another direction, and related to the domain of the biological microcosm. A geneticist wanted to know: 'How does the brain work? How do thought, memory, and biological data storage operate?', and 'Is logic inherited or acquired; what is the "logic" of biology?' These questions are already too general, too vague, to receive concrete answers, and so is the central question: 'How can cancer be cured?'

On the other hand, many scientists would be interested in the future prospects for microbiology. Perhaps mankind may one day be in a position deliberately to modify its environment—not only on Earth—for its own purposes, without destroying it at the same time. Despite the meteoric development of solid-state physics, plasma physics, astrophysics and high-energy physics during recent decades, even many physicists today have a prejudice that the 'future of science', as measured by the requirements of future generations which science must help to satisfy, lies not in physics but in microbiology.

Is physics, at least in many of its branches, becoming remote from human requirements? Is it indeed 'finished', in the sense that the most important natural laws have been established, that new knowledge is only to be expected at extremely high energies (the physics of elementary particles) or at great distances (cosmology), and thus in contexts which hardly have meaning any more for Earthly life? Is physics therefore becoming uninteresting, less important, to mankind?

To say 'yes' to this question would be to betray a naive prejudice. It would not be the first time physicists had thought that 'physics' was finished. When the young Max Planck enquired of a professor of theoretical physics in Munich about the outlook for a study of physics, he was advised against it on the grounds that everything of importance had already been discovered. But the fact is that in many branches of modern physics, either phenomena are being actively 'sought out' for study, or else the problems are almost overwhelming in their numbers, and research has come up against the barriers of unsolved problems for whose solution radically new conceptual approaches will be needed. Whether a new scientific revolution, comparable to that of the first two decades of this century, may nevertheless be in the offing, is widely doubted, but this remains pure speculation.

It may well appear reasonable to reflect on the form and content of cosmic messages and to arrive at possible drafts. But if these remain untestable, the reflections will swiftly and without interruption develop into anthropomorphic wishful thinking, not far removed from the fairy-tale world of some science fiction stories.

The easiest way to take the measure of our own bias is to put ourselves in the position of the supercivilization we are questioning. At the 1971 Byurakan Conference Frank Drake asked the question: 'How many bits does it take to transmit $E = mc^2$ [Einstein's formula for the energy E equivalent to a mass m]?' (c is the speed of light.) In the course of the same discussion Morrison gave his answer in the following story: 'In the year 1600 A.D., say, there you are studying the Fibonacci series [a set of numbers with special characteristics], and an angel appears to you and writes $E = mc^2$ in your notebook. You can't even ask what it means. Does that help you much?'[10]

Many scientific discoveries which were ahead of their time were simply ignored by contemporary scientists because they could not understand their significance. (This is still true today, for it is usually only a few specialists who concern themselves with radical ideas.) So too the sixteenth-century monk will at most shake his head or fall to mystical worshipping of the angel if the latter writes a twentieth-century formula in his notebook. The intellectual distance comprising three centuries of historical scientific thought is not such an easy leap. To take a further example from the present day: it is not much use in itself to steal the circuit diagrams of a complex electronic machine or the blueprints of a military aircraft if the technical know-how, the tools and the manufacturing precision, and so on, are not already more or less fully provided.

Many 'questions to a supercivilization' may seem equally dubious. But, at least for information in the field of physics and technology, I believe that communication is still possible, if the superior civilization adapts to the level of the questioner. It is true that we could not make the monk of the late Middle Ages understand Einstein's energy–mass relation, but we could certainly answer many interesting questions for the Greek philosopher Democritus or for Isaac Newton. On the other hand, psycho-historical statements will only be communicable with difficulty on account of the cultural difference to be assumed—or only when there is by accident a great similarity between civilizations. It is roughly the same as trying to explain the basis of our civilization to a colony of termites.

Independently of this, a civilization will rarely have the opportunity to turn to a superior society with questions of concern. The most probable occurrence is that artificial signals will suddenly be received, perhaps a year-long series of modulated electromagnetic waves, with no possibility of asking questions. In the next section, therefore, this type of interstellar message will be more closely analysed.

A computer program from Andromeda

A solution to the difficulties of transferring directly useful information between two civilizations—especially when astronomical distances render further interrogation pointless—may lie in an attempt to transmit the *total knowledge* of a civilization, in answer, so to speak, to the all-embracing question: 'What do you know?' This may then contain the answers to questions which might be put by possible listeners. Given the large time intervals demanded by a realistic interstellar exchange of information, it seems likely that an artificial signal may in fact be sent *continuously* over a period which may even be comparable to the life span of the civilization itself. In order to transmit the formula $E = mc^2$ without any assumed previous knowledge, the whole of the scientific thinking of several centuries may need to be 'unloaded' on to the addressee. To do this, we would 'only' need to transmit in the form of a message the contents of, say, all the scientific and philosophical books in all the libraries on Earth.

Let us return to an example on the Earth. Philip Morrison estimates the total amount of information contained in the culture of ancient Greece at 10^{11} or 10^{12} bits.[11] He arrives at this estimate by a rough calculation of the number of photographs which would be required to describe Greek architecture, climate, pottery, and so on. With a transmitting channel sending 10^9 bits per second (and this rate can be greatly exceeded by multi-channel transmission) the information contained in Greek culture can be sent over the transmitter in 100 seconds. Today we would face the problem that new knowledge is continually arising in our culture. Currently we are producing all the time more information than all the programmers in the world would be able to process simultaneously for continuous cosmic transmission.

It has therefore been suggested that we should not 'pour' whole libraries uncritically into the transmitting telescopes, but send carefully selected items: *hierarchical* information. Here, 'hierarchical' simply means that later information is built up on previously given information according to teaching principles. Lesson by lesson, the essential character of human existence can then be conveyed to the unknown extraterrestrial civilizations by means of examples.

The only attempt so far to solve the difficult problem of devising a new language expressed by logical symbols, such that even complex relationships, ethics, morals and historical events could be conveyed to someone who knew no terrestrial language, was made by the Dutch mathematician Hans Freudenthal in the 1950s (see 'Lincos—an interstellar language?', page 120).

An even more attractive idea, it seems to me, is that of sending not just hierarchically structured knowledge, like a sort of university course, but a computer program with built-in learning capacity. What I mean is a transmission in which the receiver is given instructions for building a programmed machine. In the early 1960s, Fred Hoyle and the television writer John Elliot published their science fiction story *A for Andromeda*.[1][2] This book, today regarded as a classic of science fiction, develops the idea of an artificial radio signal from the depths of space, in a way which expresses the deeper psychological problems, the anxieties of mankind, fears of possible hostile consequences of a contact with extraterrestrials . . .

This is the story. At a steerable radio telescope which has just been built in the North of England (we might imagine the metal dish of Bonn-Effelsberg or the Jodrell Bank telescope to have been transplanted there) the technicians receive mysterious radio signals, even before it is officially put into operation. 'In among the crackle and whistles and hiss from the speaker came a faint single note, broken but always continuing.'

Dr John Fleming, a young astronomer, soon realizes that the phenomenon is an artificial message from the direction of the galaxy M31, the Andromeda Nebula—2.5 million light-years away. To the Minister's enquiry, 'You can decipher it?', Fleming gives the irritable reply: 'For heaven's sake! Do you think the cosmos is populated by Boy Scouts sending morse code?'

But a little later, Fleming is able to announce to a Cabinet meeting: 'It's a computer program.' Further: 'It's in three sections. The first part is a design—or rather, it's a mathematical requirement which can be interpreted as a design. The second part is the program proper, the order code as we call it. The third and last part is data—information sent for the machine to work on.'

Finally, Fleming succeeds in getting the largest military computer in the country placed at his disposal for a run of the program, and makes the necessary adaptations. The program is read in, and promptly begins an interrogation of the scientists at the observatory. It asks for information about the biochemistry of terrestrial life, first offering an energy model for the electron in the hydrogen atom. Fleming gives an affirmative answer. Then the computer pours out more columns of figures, including symbols representing the carbon atom. Fleming conjectures: 'He's asking us what form of life we belong to. All these other figures are other possible ways of making living creatures. But we don't know anything about them, because all life on this earth is based on the carbon atom.' After this come complex protein molecules, haemoglobin, chromosome structures . . . At every level of biochemical complexity, the program offers several possibilities, and Fleming's colleagues type in the 'correct' terrestrial version in answer. In this way the program acquaints itself with the basic pattern of the origin of life on Earth, terrestrial biosynthesis. Under computer control, the unthinkable then comes to pass: an organism is synthesized, a sort of cyclops. Fleming concludes:

> [The intelligence] puts out a message that can be picked up and interpreted and acted upon by other intelligences. The technique we use doesn't matter, just as it doesn't matter what make of radio set you buy—you get the same programmes. What matters is, we accept their programme: a program which uses arithmetical logic to adapt itself to our conditions, or any other conditions for that matter. It knows the bases of life: it finds out which ours is. It finds out how our brains work, how our bodies are built, how we get our information—we tell it about our nervous system and our sensory organs. So then it makes a creature with a body and a sensory organ—an eye.

Later, in a second run after the dramatic murder of a woman scientist by the cyclops, a human being is produced: a girl with the features of the murdered one. Shortly afterwards the computer kills the cyclops. The girl, Andromeda, once she has learned to speak, comments: 'We were only eliminating unwanted material.' As yet, no one suspects that Andromeda—the medium of a super-powerful extraterrestrial intelligence—is seeking to sieze for herself power over the Earth . . .

It is no accident that the 'Andromeda' story was written shortly after the conclusion of the Ozma project (see 'Project Ozma and its successors', page 155); and yet it went much further. The story aimed to show that knowledge can be more easily transferred between two civilizations if it is conveyed in a *creative* manner. Questions can be asked, and knowledge can be built up from wrong answers. Of course, the story merely embodies a fascinating idea and hardly a realizable concept. Questions remain open which even now can not be answered, despite considerable advances in computer technology. Least convincing is the unambiguity of the Andromeda message. It is not made at all clear how—even allowing for flashes of genius—it was so quickly discovered to be a plan for a computer; and where did the computer get its encyclopaedic knowledge of all possible forms of life? Was this information included in the transmission from the start? This point is completely skipped over. According to what we know today about the enormous variety of life-forms, the amount of information necessary to transmit this knowledge seems far too large for the duration of the message from the Andromeda Galaxy.

Computers which learn from their mistakes

Compared with 'Andromeda', the scientific realities of today are very modest, almost antediluvian. For some time certain branches of cybernetics have been working on the development of machines capable of learning. An example is seen in programs for automatic pattern recognition (see 'Pictograms and the computer: automatic image recognition', page 132): a particular recognition process is repeated several times, and at each step the knowledge last obtained is made the new basis for proceeding. There are also computer programs which can learn to win at certain board games in which two players make moves alternatively. After several years of effort, for example, a program was written for a game known as *Kalah*. The computer scientist John McCarthy at Stanford University writes: 'The computer now plays Kalah better than any human players that we have been able to find.'[13] But humans, too, have improved their ability to play Kalah by watching the computer playing for a while. Computer experts have found from an analysis of this computer game that the player who has the first move can be the winner—a fact not known before.

A computer program developed for the more difficult game of *draughts* (checkers) played at least well enough in a match against the American champion to draw one of six games. But once the champion had worked out *how* the program played, he beat it almost every time.

Computers master board games better, the fewer decision routines they have to go through in order to find the best move—decision chains of the type: 'If I do this, he will do that, so I'll do this, then he'll answer with that, and I'll lose; but if I move like this, then . . . ', and so on. The greater the variety of moves permitted at each point of the game, the worse showing the computer makes, as it very soon becomes impossible with the memory storage systems available today (and from the standpoint of strategy pointless too) to run through the mushrooming number of possible variants which fall within the rules. Chess, and especially the Chinese game *Go*, thus present the greatest difficulties to computers today, since strategies—chains of decisions whose ramifications extend over many moves— are difficult to program. Although the most time and mental effort has been invested in chess programming—the best in existence today comes from the American researcher Greenblatt at Massachusetts Institute of Technology— these can still be beaten by talented amateur chess players: in 1975 the then West German Minister of Research Matthöfer checkmated an American chess program in less than half an hour.

Much work has been devoted to the branch of computer research concerned with the interaction between computers and their environment.

In his doctoral work at MIT, Terry Winograd made a practical study of the problem of developing a simple computer program with learning capacity. Winograd's intention was to write the program in such a way that it could hold a meaningful conversation with someone in English. In recent years this work 'has attracted considerable attention as the most dramatic and important achievement so far in artificial intelligence', according to Miranda Robertson, an associate editor of the British scientific journal *Nature*.[14]

The program deals with a very limited world, in which there are merely several coloured building blocks, which can lie beside or upon one another, inside or outside a box. Only this world can be 'discussed' with the program. Let us first see what it can do: here is the course of a typical conversation.[15]

Via a teleprinter the program is asked: 'What does the box contain?'

The machine answers, 'The blue pyramid and the blue box.'

'How many blocks are not in the box?'

'Four of them,' says the program.

'Is at least one of them narrower than the one which I told you to pick up?'

'Yes, the red cube.'

The program correctly answers questions as to how the objects known to it are arranged. But it can also carry out orders and learn new concepts:

'Is there a steeple?' asks the operator.

'Sorry, I don't know the word "steeple",' says the machine.

The operator elucidates:

'A steeple is a stack which contains two green cubes and a pyramid, for example.'

'I understand,' replies the machine.

'Are there any steeples now?'

'No,' replies the machine.

'Build one,' commands the operator, and the machine carries out the order. It stacks blocks on one another, and even builds bridges—though not in reality: Winograd never connected a mechanical arm to the system. It was sufficient to demonstrate that the program 'knew' how it had to move the blocks to carry out the required command. But Winograd's machine also answers questions about *when* and *why* it does what it does: a large part of the subsidiary programming is concerned with self-diagnosis. To the question:

'When did you pick up the pyramid?', the machine responds:

'While I was stacking up the red cube, the large red block and the large green cube.'

'Why?'

'To get rid of it,' answers the machine.

'Why did you do that?'

'To clear off the red cube.'

'Why did you clear off that cube?'

'To put it on the large green cube.'

If it is asked 'Why?' over and over, the program will give a series of mechanical reasons, until finally only one answer remains: 'Because you asked me to.' Despite the apparent simplicity of the building-block world, the programming effort was enormous. Several million bits had to be incorporated in Winograd's computer code in order to achieve intelligent behaviour even of so simple a form. This fact demonstrates the surprisingly large amount of information which is generally taken for granted even for simple manipulations. And even within this limited world only simple things can be discussed, even when the questions are quite closely limited to what the machine can just understand.

There was a general motive behind Winograd's computer research: to develop a program which was in a position to undertake minor programming tasks. On being given specific cues, the system should be capable of carrying out further programming and of co-ordinating the derived program unit into the existing system. This idea has long been familiar to the professional science fiction consumer. In Robert A. Heinlein's novel *The Moon is a Harsh Mistress* (1966), it is developed in detail for a computer in the lunar colony. In the struggle for the independence of the Moon colony, Heinlein's lunar computer develops 'self-consciousness', or intelligent and articulate behaviour, and assists a man in the defence of the Moon colony by programming itself to direct the optimum use of weapon systems. Winograd's computer system consists of a series of programs, some of which control others, and which also give information as to what the program knows, and how it would react to interference. In Winograd's words: 'This way, at least we can always rely on our computers saying what they think.'[14]

These internal control programs are essential to the learning capacity of the system. In a later stage of development they are even expected to take over the task of correcting programming errors independently. In the opinion of the MIT scientists, the processes of error correction are an important aspect of 'creative thinking'. Here we can already recognize an analogy with human behaviour: in modern learning psychology there is now a trend towards regarding the making of errors as an important component of the learning process. The errors made by the learner are positively assessed and not suppressed by punishment. Experimental teams in the United States have shown that specific learning goals are far more quickly attained by an appreciative attitude towards errors than by the usual method of penalizing by low marks. I will give here another example of the way in which self-analysis and self-correction operate in Winograd's computer system. The operator commands the system through the teleprinter to build a tower from three blocks. From its standard programming the machine 'knows' how to carry out the individual step necessary for this, that is, how to put a given block A on another block B. But to stack three blocks—A, B and C—on one another, this program unit must be executed twice, as a man would see intuitively at once: first to put B on C, and then again to put A on B. Then A is on top of B and B on C, and there is your tower. Winograd's machine reacts differently: as the first step, it puts A on B and forgets the action. Now it tries to put B on C. But block B is now covered by block A. The program can only move one block at a time; so it takes A down from B, and is now able to put B on C. So it first put A on B, and two steps later, B on C, just as the operator commanded; but it has not completed the tower. The system only finds this out by making a final check, and now falls to meditating.

How can the system avoid this mistake? How can it correct its error while still building the tower? To find out exactly what it 'thought' at each step, the machine goes through the process again from the beginning. But this time it uses an additional program, which precisely analyses the situation at each step of the operation, and provides a commentary on the action. Since the principal program itself 'reads' the commentary, the program 'knows' what has just happened. As soon as the subsidiary program once more begins the false step of removing A from B, it is immediately halted. Now the 'error analyser' takes command, and minutely scrutinizes the subsidiary steps which led to this false move. At the end of the investigation it finds out the nature of the error, and sets up an appropriately corrected course of action. The results of the error analysis are stored, and next time the system will make use of this 'experience'. It will not make the same mistake again. The system has learned something new.

What is intuitively clear to a man, the computer must learn by electronic hard work. The general principle behind Winograd's programs is that the system should know what it is doing and be in a position to learn from its own mistakes. Winograd and other computer experts have taken the first step towards giving 'self-consciousness' to a machine. Modest as the achievement may be, it has taken computer technology into a new era in the development of artificial intelligence. Finally, the mechanization of self-analysis also brings the opportunity of gaining more insight into the course of intellectual processes. And the goal of the automatically self-programming computer now seems, at least in principle, to be attainable.

These beginnings of 'rational' artificial intelligence are modest, and currently throw more light on the drastic difficulties of self-conscious cybernetic systems. Whether the complexity of the human brain can ever be attained or even exceeded is something which cannot yet be decided. But it seems at least possible that within 50 or 100 years specialists in a wide variety of fields will converse with highly developed computers and write themselves programs, without any familiarity with the complex art of programming, which will have been mastered by computers capable of learning and self-programming. But for the moment, Hoyle's 'Andromeda' computer remains science fiction.

Lincos—an interstellar language?

An important problem in interstellar communication is the *language* in which the messages are expressed. Whether the intention is to send the plans for constructing a 'talking' computer into the Galaxy or only a one-way message, the news must be coded: graphically, in pictorial form, or non-graphically, in a 'text'.

In fact, in 1960 a mathematician actually undertook the heroic task of devising such a language in concrete form. Hans Freudenthal published a book in that year with the title *Lincos—Design of a Language for Cosmic Intercourse, Part I*, in which he constructs a complete body of language on specific logical principles. *Lincos*, an abbreviation for *lingua cosmica*, is an artificial and purely schematic language, not designed to be spoken. In contrast to natural languages, *Lincos* therefore avoids all irregularities and exceptions. This logical levelling is designed to make understanding easier for an addressee who knows no terrestrial languages. 'My purpose is to design a language that can be understood by a person not acquainted with any of our natural languages or even their syntactical structures,' says Freudenthal in the introduction to his book. 'The messages communicated by means of this language will contain not only mathematics, but in principle the whole bulk of our knowledge.'[17]

With the *Lincos* language, built up on pure logic, Freudenthal believes that he can also introduce and make comprehensible non-mathematical concepts. New concepts will first be introduced in a very intuitive manner. For this purpose he falls back on the elementary form of a dialogue. How do two people with no common language communicate? They use sign language and point to objects. Similarly, *Lincos* introduces new concepts via their usage in dialogues between two fictitious 'people'.

Lincos is a sort of correspondence course: it consists of chapters and sections built hierarchically one on another. This obvious structure is designed to make understanding as easy as possible. For the interpretation of the language cannot be found by logic within the linguistic system itself. This is a problem of fundamental, if inescapable and obvious significance. It embodies a basic property of 'an isolated system of symbols', such as an interstellar message always represents. No isolated message can carry its own interpretation, merely within the symbols of the message itself. In other words, an interpretation cannot be extracted from the actual form of the message.

Many critics rightly see a fundamental limitation to all attempts to develop languages for cosmic messages à la *Lincos* in the incapacity to make a finite message self-interpreting. The situation is simpler for radio transmission: the physics and technology must be well known to both transmitter and receiver. Here there is quite solid common ground. In practice the receiver has no choice at first but to guess what objects are referred to in the message. The more frequently these objects are presented in different contexts, the more tests of consistency he can undertake, and thus continuously improve on his first interpretation. A *Lincos* signal is designed to make this guessing game as easy as possible. The transmission will provide enough examples and contexts for a concept or a statement for us to expect the receiver himself to be able to generalize the quasi-general definitions in the text. But nevertheless Freudenthal himself does not exclude the possibility of misunderstandings. But in spite of our efforts even

LIBRARY

intelligent receivers might interpret our messages as physical phenomena or as music of the spheres.'[17]

The *Lincos* course begins with the simplest concepts of logic and mathematics; then the vocabulary is built up lesson by lesson, and dialogues illustrate by what rules and in what contexts new concepts are used. The first lessons might take the following form (the actual message on the left, its meaning on the right):

MESSAGE	MEANING
	Chapter I Mathematics
- -- ---	Lesson 1: Natural numbers
and so on	Lesson 2: Code numbers
- = 1	The sign of identity (=)
-- = 2	and symbols for the natural
--- = 3	numbers are introduced

The symbols for the 'equals' sign and numbers introduced in this lesson will perhaps not be immediately understandable, but Freudenthal thinks that the repeated use of the symbols in various contexts in the following lessons will remove such uncertainties; for instance, in the lesson below, where simple calculations are introduced:

MESSAGE	MEANING
	Lesson 3: Addition
$1 + 1 = 2$	Simple 'sums'
$1 + 2 = 3$	are illustrated
$1 + 3 = 4$	
and so on	

More complex mathematical laws could be introduced fairly soon in similar lessons: negative numbers, primes, fractions, real numbers, complex numbers, statements of number theory, group theory, algebra; and also concepts such as: 'is different from', 'decreases', 'increases', up to more difficult statements like: 'it is true that' and 'it is not true that'. In Freudenthal's plan the fundamentals of mathematics are followed by some chapters on the concept of time, human behaviour, and finally physics (space, motion, matter).

In the lessons on human behaviour 'people' are introduced by means of dialogue: person A talks to person B. Indeed, Freudenthal hopes that in the linear order of transmission, successive sentences will be recognizable as conversational elements like questions and answers. Actions are judged as being either 'good' or 'bad'. These two fictitious people now introduce further knowledge. At first they converse about mathematics. One person asks a question, the second person answers it, and the first says whether he considers the answer right or wrong. Here is an example:

MESSAGE		MEANING
		Chapter: Human behaviour
		Lesson: Mathematical ability
A.	What is 2 + 3?	The words 'correct' and
B.	2 + 3 = 5	'incorrect' are introduced
A.	Correct	into the dialogue
A.	What is 10 × 10?	
B.	20	
A.	Incorrect	
B.	100	
A.	Correct	

Further in the development of Freudenthal's *Lincos* course come such concepts as 'let us assume that', 'roughly', 'hence', 'instead of.', 'why' and 'because'. And also 'almost', 'much', 'little', 'now', 'necessary', 'possible', 'to know', 'to understand', 'to think', and 'to mean'. Men are more precisely defined by 'age', 'existence', 'birth' and 'death'. With an increasing vocabulary, more and more complex contexts of human behaviour are covered by the *Lincos* program, lesson by lesson. We may also imagine *Lincos* supplemented by pictorial illustration, like the Arecibo message (see page 138).

Perhaps the reader will regard this program as a fairy-tale. And, in fact, apart from Freudenthal's logical and constructive achievement, *Lincos* should not be overestimated. There are no really plausible principles by which *Lincos* can be applied in practice, so that the 'success' of a *Lincos* transmission is indeed extremely doubtful. It is hardly realistic to expect radioastronomers to fall back on *Lincos* in the composition of their messages. In addition, the receiver must pick up all the lessons: this is certainly one of the weaknesses of hierarchical messages. It can easily happen, on account of technical faults, that important sections of the program are missed and not recorded. But this danger can be avoided: the transmitter must merely repeat the message often enough.

But Freudenthal's *Lingua Cosmica* has also been criticized for other reasons. It has seemed to many people almost impossible to teach such a relatively complicated language when the receiver has no possibility of asking questions back. Indeed, attempts have been made in the Soviet Union to instruct students by means of *Lincos*, and N. T. Petrovich considers that the first experiments have 'not yielded negative results'.[18] But he could also not say whether *Lincos* had been very satisfactory; however, he thinks that a simpler language system than *Lincos* itself can be constructed for interstellar communication. Nevertheless, Freudenthal's *Lincos* still remains the only consistent attempt to construct an interstellar language, at present certainly no more than a heroic attempt to offer a learnable language to interstellar intelligences. There are many conceivable situations in which *Lincos* would be totally unsuited to establishing contact, and where all attempts at contact on Earthly lines would fail (see 'Unrealizable contacts: Rama, Solaris and The Invincible', page 198).

The problem of decoding in cosmic linguistics

'*Arecibo, Puerto Rico, 20 March 2028 (AP)*: For two days, radioastronomers at the world's largest radio telescope have been picking up an artificial signal from outer space. According to a TASS report, the discovery has been confirmed by scientists at the Soviet radio telescope in the Ukraine. Early computer analyses show that the source of the signals lies in the same direction as the globular star-cluster M 13, a star-cluster in our own Galaxy but 24 000 light-years away from the Earth. At a press conference carried live on the World Vision network yesterday, the telescope's research director dismissed the possibility that the signals might come from the Earth itself. It was also most unlikely that the message might be coming from a hitherto unknown, natural astronomical source. Some decades ago, astronomers had several times drawn attention to themselves by erroneously attributing signals from Earth to extraterrestrials. It is already clear beyond doubt that the signal, which continues to be received, has an artificial origin—but scientists are still completely in the dark as regards the content of this message from the stars. It appears to be based on a complex type of code beyond the capacity of computers designed for cosmic linguistics. According to linguistics expert and White House spokesperson for military intelligence, Professor Isaac Colgate, the message had lost a good deal of clarity during the several thousand years in transit between the stars . . . '

Associated Press reports like the science fiction example above might serve for several weeks in the year 2028 to stir up excitement in the public media.

But the loss of information 'en route' can make the actual message incomprehensible. The restoration of messages partially destroyed in transit is one of the central problems of communication theory. But how alien—to our conception—will the reality be which lies hidden behind an extraterrestrial message? Here we must stress that there is a great difference between 'understanding' and 'comprehending' the 'real content'. Understanding rests on our ability to supply missing parts of a message or to anticipate new ones. A fuller comprehension, on the other hand, allows us to visualize concepts pictorially and place them in the context of our own reality. Not everything that we can understand is easy to comprehend. In the cosmic exchange of information we shall have to content ourselves, where complex signals are concerned, with understanding these signals, even though the background reality—beyond the transmitted information—may remain totally concealed from us.

The decoding programs used today by astronomers to process the observational data of individual stars, gas clouds or distant galaxies have only a limited application to artificial signals from other civilizations. Two kinds of artificial extraterrestrial signals must be distinguished: those that are sent out for the *purpose* of establishing interstellar contact, and those that merely leave the sphere of the civilization *accidentally*. The latter might be electromagnetic waves produced by radio, television and radar systems; but they could also be the traces of a civilization, left behind on their planet by a dying race.

Since the beginning of the 1930s the Earth has been an involuntary sender of interstellar signals. Up to now, these have spread out over a roughly spherical area of 50 light-years' radius. To overhear accidental signals, to eavesdrop, to 'listen in', so to speak, and to use terrestrial telescopes for cosmic 'bugging', is certainly more difficult. The signals will be very weak, since they are emitted isotropically—as is the case on the Earth; besides, we can guess at few of the characteristics of these signals beforehand and thus cannot design our receivers for particular types of signals. The question is, have we any realistic chance of decoding interstellar messages? Most of the suggestions which have been made may be so closely tied to terrestrial patterns of thought and our species' cultural, sociological and biological development—so anthropomorphic—that they can have no meaningful application to interstellar signals. So a cosmic linguist has little cause for optimism: even today we have not deciphered all the writings left behind on Earth by earlier advanced cultures. If we come accidentally on the 'internal' signals of a civilization, we must be prepared in principle for a complexity at least on a par with that reflected in human language and literature. Language mirrors our physical environment—the Earth, the Solar System, the Universe—but also the life-form in its biology, psyche, intellect and society.

Carl Sagan, too, is sceptical: 'European scholars spent more than a century in entirely erroneous attempts to decode Egyptian hieroglyphics before the discovery of the Rosetta Stone [1799] and the brilliant attack on its translation by Young and Champollion. Some ancient languages, such as the glyphs of Easter Island, the writings of the Mayas, and some varieties of Cretan script, remain completely undecoded to the present time. Yet they were languages of human beings like ourselves, with common biological instincts and encodings, and distant from us in time by only a few hundreds to a few thousands of years. How can we expect that a civilization vastly more advanced than we, and based entirely upon different biological principles, could ever send a message we could understand?'[19]

It will probably be easiest for us to establish the physical environment in which an extraterrestrial civilization has developed its language—that is, the class of its central star and the characteristics of its planetary system. The laws of physics are universally valid; given the spectral type of its star, it may even be possible, in certain circumstances, to deduce the properties of its biological life-form.

For decoding a language itself we should, of course, make use of the experience gathered in the decipherment of ancient languages. Here we have a case where we receive artificial signals from the earlier history of human culture which have not been purposely coded. But we are involuntary eavesdroppers and spies; although the messages come from our own ancestors, the complexity of the earlier natural languages has hitherto prevented successful translation in a number of cases. Even in the cases that have been solved, the 'key' was often found only by accident or by a daring guess. The Egyptian hieroglyphics were solved by bilingual—that is, already translated—inscriptions; the language of the Hittites was 'broken' in 1938 by F. Hrozny after he had made the daring postulate that it was related to the Indo-European languages. As a rule, the special solutions applied in such brilliant work cannot be used for the unsolved cases, for they depend to an extreme on human intuition, on prior knowledge of the particular epoch (wars, trade, etc.) and on fortuitous written finds. At all events, a systematic linguistic procedure cannot be derived from them.

How to proceed then? Without further prior knowledge of the civilization this would imply the decoding of a (linguistically) practically arbitrary text—an almost hopeless undertaking. Today, at least, the idea of decoding a message which has merely been overheard still seems hopeless. For a message intentionally addressed to the peoples of the Universe, there is at least a chance of successful decoding. The possibilities for understanding interstellar messages must be given a most pessimistic verdict.

Besides this, the message may be so garbled (possibly artificially and therefore deliberately) that it can only be 'de-jammed' by means of a decoding key. In simple cases a code can, of course, be broken without this information. The American Secret Service proved this when they deciphered the Japanese secret code during the Second World War. The British Secret Service, too, deciphered the code of the German navy with the aid of the first electronic computer. The decoding is the more difficult, the more information is involved in the encoding process as compared to the information content of the message itself. It may even be so thoroughly encoded that it is in principle undecodable. C. E. Shannon, the founder of modern information theory, shows in his fundamental book, *The Mathematical Theory of Communication*, that a text can be so encoded by a code of comparable information content that there is no conceivable method which will retranslate it.[20] The following code may be selected as an example: Number the individual letters of a message from beginning to end with the natural numbers 1, 2, 3 . . . and write each number on a card. If the cards are now shuffled and the letters arranged according to the newly produced sequence of numbers—the code—the message is coded. Unless the code is known, decoding is in fact impossible, for the code resides in a random order of numbers. The frequency of letters characteristic of a language is not even changed in the process. By adding the appropriate letters, this language characteristic can also be smoothed out.

The decoding team for cosmic messages at the radio observatory now faces the task of deducing from the portion of the message which has been received the content of the gaps. The sentence: 'Einstein lived in the twentieth chapter', is in itself not understandable unless we amplify it by adding the word 'century' to the partial sentence 'Einstein lived in the twentieth . . . '. But in the pulsar message ' . . . yes–no–yes–no . . . ' any number of gaps may appear without presenting problems; it can always be made complete, provided that at least the 'yes–no' period is visible in the portion which has been received.

In the decoding process a message will usually admit of a number of interpretations, depending on the extent of the gaps in the text and the complexity of the code. The preferred interpretation must then be taken as the one which best permits the message to be understood. This presupposes that a sort of dictionary of 'words' and a corresponding grammar can be constructed, with whose aid the gaps in the text can be filled. Aside from this formal decoding, however, the actual aim is the translation of the message, by which the elements of the signal can be given a relation to words in a language we understand. Only when one of several possible interpretations is obviously 'more sensible' than the others—as is to be expected from a call signal preceding other intentional messages—will the message have a *meaning* in the linguistic sense.[21] To have a meaning is perhaps the most important property of a message. If there is no success in providing a sufficiently understandable interpretation, the message, regarded in itself, is rather uninteresting—'meaningless'. This is the case for continuous pulsar transmissions ('yes–no–yes–no . . . ') and for nearly trivial messages. For these all possible interpretations are equally understandable.

Cosmic pictograms—mankind in outline

Early in 1959 a meeting of specialists was held at the newly built National Radio Astronomy Observatory in Green Bank, at which Lloyd V. Berkner, then Director of the Observatory, informed the prominent astronomer Otto Struve of the envisaged project to search for extraterrestrial signals on the 21-centimetre hydrogen line. This idea had been suggested to Berkner a little earlier by the then young astronomer, Frank Drake. His project, begun on 8 April 1960 and concluded after 150 hours of observing the stars Epsilon Eridani and Tau Ceti, entered the history of the search for signals from space as the pioneering 'Project Ozma' discussed in a later section (see page 155).

After the Green Bank conference, Drake sent an unusual message to all the participants, as Walter Sullivan, scientific correspondent of the *New York Times*, tells in his book *We are Not Alone*.[8] Drake intended it as an example of how the first message from another civilization to the inhabitants might look: the message was composed in binary code—that is, using only the symbols 0 and 1—and contained 1271 ones and zeroes, corresponding to 1271 bits.

Most of Drake's colleagues decoded the data only in part. Bernard M. Oliver of the Hewlett Packard Corporation (the firm whose present fame rests on pocket electronic calculators) 'cracked' the binary telegram in the following way: first, and most importantly, he examined the number 1271: it is the product of two prime numbers, 31 and 41. This offers the possibility of dividing the continuous 'line of text' of the message into either 31 rows of 41 characters or 41 rows of 31 characters. The next step is to fill this rectangle with a black-and-white mosaic: 1 = black, 0 = empty = white. The reader may like to try out the two possibilities on graph paper—it will be clear that only one of the two interpretations 'makes sense'.

Here is a 'translation' of the diagram (see Figure 32). It shows three inhabitants, two parents and a child, similar to the two-legged mammals on the Earth. The circle in the upper left-hand corner is meant to represent a sun, with several planets beneath it. The 'man'—an organ like a penis is suggested—points to the fourth planet. The third planet—it is clear to us—is covered with water and populated by fish: a wavy line adjoins the planetary symbol, with a picture of a fish below it. On the upper edge, beside the sun, three atoms are represented by nuclei and the number of encircling shell electrons: hydrogen, carbon and oxygen—the chemical basis of their life, just as on Earth. A scale to the right of the inhabitants gives the number 11 in binary code: 11 units of length? The only length common to sender and receiver is clearly the wavelength of 21 centimetres, with which it was chosen to send the message. Thus, the creatures are $11 \times 21 = 2.31$ metres tall, also a measure of the surface gravity of the planet that they inhabit.

Drake found that hardly anyone deciphered the full content of the picture. 'Indeed if such a picture is given to experts, they identify those parts of the message associated with their specialty but normally fail completely on the part that they are not expert in.'[22] Drake has since devised an improved and considerably more compact version of this pictogram, using only 551 bits. In many ways it exposes the disadvantages of such pictograms—they are bound in the extreme to terrestrial ways of seeing—rather than their suitability as nearly self-decoding, anti-cryptographic cosmic messages. The code is the same as before: the receiver must realize that he has been presented with a two-dimensional picture, and to find the lengths of its sides he must seek to resolve the number of bits into prime numbers: $551 = 29 \times 19$. This resolution excludes a three-dimensional picture: 551 cannot be broken down into three (non-trivial) factors. Tests will show that the correct interpretation lies in a picture consisting of 29 rows of 19 units, since only this version looks 'sensible'. Again, the figure of a primate is most easily recognized: a creature with short legs and a sort of lump on his head—a sign of a planet with a higher gravity? At the left-hand edge of the picture astronomers will recognize a diagram of a sun and its planetary system: four small planets, a middle-sized one, two large ones, a medium-large one and a small body—a planetary system like the Sun's. On the upper right is the section for chemists: a carbon atom (nucleus, two inner and four outer electrons), and an

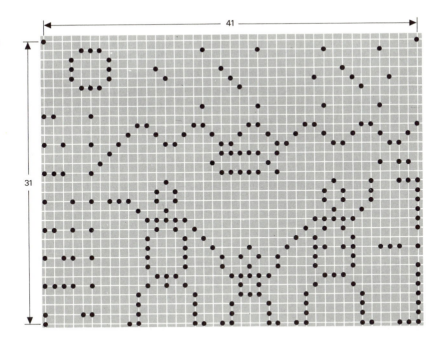

Figure 32 A message to the stars? Solution to the first cosmogram by Frank Drake, with the original as sent.

oxygen atom (nucleus, two inner and six outer electrons). The group of symbols beneath is new compared to the earlier version, and gives the key to the rest of the information in the picture. This group presents the numbers 1, 2, 3, 4 and 5 in binary code, but with an additional '1' if the binary notation contains an even number of ones, as for instance in the case of 3 (11 in binary code) and 5 (101). These are now written 11-1 and 101-1, so that the number of ones is always odd. This use of 'parity bits' is normal practice with computers, where it is used as a check on the correct transmission of data. But for its inventor, Drake, this was not the reason for using this device here. If all groups of symbols with odd numbers of ones denote figures, it follows that the word beneath the creature is not a figure: perhaps it gives the name of the creature, in which case this will be confirmed in later messages. Then, once it has been made clear to the receiver which are numbers and which are words, it becomes apparent that the diagonal line connects a number with the creature. There are numbers beside some of the planets—perhaps they denote how many creatures live on those planets: 5, 2000 and about 4000 million. If this denotes population, we could deduce that 4000 million creatures live on the fourth planet, a colony of 2000 on planet No. 3, and some kind of scientific expedition of 5 creatures on planet No. 2. The scale of size to the right gives the binary number 31 (= 11111). If the message were received at a wavelength of 10 centimetres, this would indicate that the creature is 3.1 metres tall.

It is a typical property of this pictogram that it conveys much more information than the number of bits. But this is a consequence of the fact that the information is only given in the roughest outline. The oxygen atom is not really 'described' in its physical complexity. We merely guess that it could be an atom and compare it with a number of possible interpretations (and in this case a catalogue of all known atoms), and select the one which fits the symbols best.

Drake's purpose was to use these pictograms to offer conceivable examples of interstellar communications and their information content, and to give this in a code which does not conceal the information but is designed to be deciphered; messages which can be fully interpreted and thus understood without any previous understanding between the participants, which render superfluous any dialogue or supplementary questions—which are in any case as good as impossible—and which are presented in the language of science, whose structures are the same throughout the whole Universe.

It is rather questionable whether Drake's pictograms actually fulfil these conditions. The problem does not seem to me to lie so much in the additional knowledge required for the interpretation of pictograms. We certainly could not build a radio receiver without at the same time having a theory of the hydrogen atom. In my opinion the difficulty lies in the two-dimensional nature of the representation, which is bound to human sight organs. Can it, for example, be understood as a two-dimensional sketch of a hydrogen atom by an intelligent gas cloud, such as that in Hoyle's science fiction novel *The Black Cloud* (1957)? And

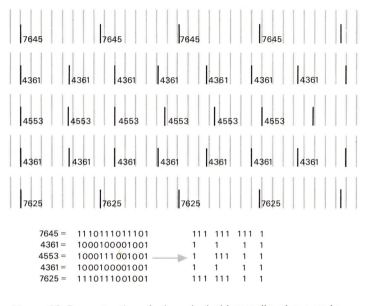

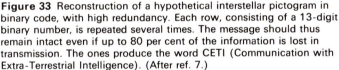

Figure 33 Reconstruction of a hypothetical interstellar pictogram in binary code, with high redundancy. Each row, consisting of a 13-digit binary number, is repeated several times. The message should thus remain intact even if up to 80 per cent of the information is lost in transmission. The ones produce the word CETI (Communication with Extra-Terrestrial Intelligence). (After ref. 7.)

the various numerical data in the picture can be understood in another way. The number of inhabitants of the fourth planet can equally well be read as three smaller numbers, with the interpretation that there are three kinds of living creatures there, each in very small numbers.[23]

A further problem is the question of the *redundancy* of the picture text—the amount of information which can go astray without garbling the essential content. Let us assume that only one of the 551 bits has been lost in the whole process of transmission. This would correspond to the excellent reproduction rate of 99.82 per cent. In spite of this, the important first step, the resolution into prime factors, at once ceases to be applicable. Now again it is an open question as to how many dimensions the message has: the number 550 has a quite different range of prime factors. But this problem can be avoided if, for example, each individual row is repeated several times (see Figure 33). In this way the length of the rows and the number of symbols per row are unambiguously defined, without the need of the device of resolution into prime factors. The message might be further guarded against transmission errors if it were also repeated several times in its entirety.

Independently of these questions it should be added that factual knowledge is here being illustrated in a manner which only developed on Earth over several hundred years, and which has its roots in our cultural history. This may well, perhaps, present no great problems with macroscopic objects like the Solar System. But a representation of living creatures as two-footed primates rather definitely requires for its correct understanding a knowledge of biological individuals so constructed. The wavy line and fish in Figure 32 today bring to mind children's picture books, rather than the first concepts we might wish to convey to another civilization. For microscopic objects like atoms, which symbolize the entire organic chemistry and biology of the life-form, I doubt whether they furnish a good example of easily comprehensible coding. The pictorial way of representing atoms as nuclei with shells and electrons as points, like the Sun and its planets, dates from the early days of quantum mechanics and from the—as we see it now—naive atomic model of the Danish physicist Niels Bohr. The whole controversy as to the 'correct' interpretation of quantum mechanics in the 1920s and 30s—the Copenhagen Interpretation—and the 'paradoxical' dualism of waves and particles merely reflects the historical conditioning inherent in scientific outlook and conceptualization. The representation of atomic structures as well as the mathematics and axioms of its logic may be quite different in other civilizations.

The recognition of an image is only possible for a life-form which possesses an organ for the perception of electromagnetic waves (Figure 34). Which frequency band is 'seen' by it is of secondary importance. The human eye is most sensitive at the wavelength at which our star, the Sun, radiates the most energy: the colour green. Bees also see in the ultraviolet, where human sight organs are already blind, and they find their way about within the same environment, though with different requirements. It is not so very unlikely that beings on other planets can also see. But can they recognize above and below, right and left, in a two-dimensional picture?

Pictograms and the computer: automatic image recognition

Despite the reservations above, pictures, either two- or three-dimensional, should be regarded as important aids to interstellar communication. A cosmic linguist cannot confine himself to a (one-dimensional) concept of comprehensibility such as one adequate, for instance, for the interpretation of language texts. Because he must be prepared for almost 'anything', the framework he chooses must be flexible enough to take into account at least the methods we know of by which real phenomena and the abstract structures built on them can be converted into communicable signals. B. V. Sukhotin, discussing these, calls especial attention to the significance of two-dimensional pictures: 'The translation from the

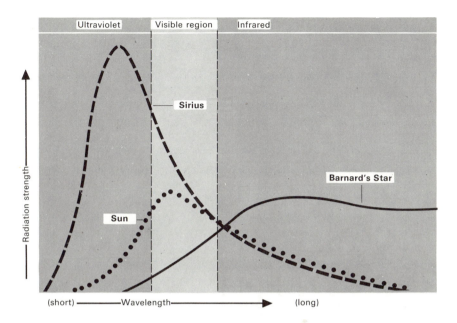

Figure 34 The light of stars has different strength at different wavelengths.

language of reality into human language and back again is naturally a complex undertaking. There is, however, one human language in particular for which this translation can be carried out without great difficulty. By this we mean the language of pictures.'[21]

On one hand, pictures as a medium of communication are very uneconomical. Only compare the sentence: 'a man is walking', with a photograph of the same process: an exact picture contains many details which are irrelevant to the essential information. And of course, there are, besides, many statements which cannot be expressed in picture language without considerable simplification, especially if abstract concepts occur in them such as, for example, the words ascribed to Edith Piaf: 'Morality is living so that living is no fun.' These examples alone show that the language of pictures will never be the only serviceable means of interstellar communication.

On the other hand, communication by pictures has the great advantage that it is easily understood, at least by beings provided with optical sense organs (see Figure 35). Among other things, understanding implies the capacity to fill in gaps—to accept fragments of information and extrapolate the rest. The more details, that is *the more information* a picture contains, the easier this is. In pictures, adjacent sections of the picture stand in especially close relationship to one another, and gaps are thus all the easier to close. With pictures the situation is

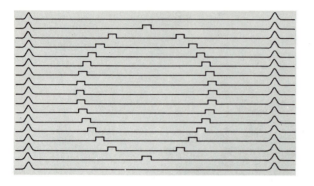

Figure 35 Representation of a circle (top) and Pythagoras's Theorem by pictograms.

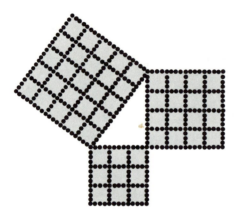

the exact opposite of that concerning the analysis of language texts: the latter are more comprehensible, as regards the filling out of gaps, *the less information* is conveyed in the text.

A computer is programmed for automatic image recognition in such a way that it reacts to especially striking associations: edges (horizontal, vertical or diagonal), patterns, colours. The straight lines occuring in each possible version of the picture can be used as a measure of the degree of comprehensibility: the more there are, the more comprehensible it is. Drake's cosmograms were also put through this process, and the computer's selection procedure found the 'correct' picture without fail. The computer program was extended to more complex pictures, in which curved lines and edges occur. The problem of interstellar communication by pictures, as presented to the cosmic linguist, is this: a linear series of signals must be transformed into an ordinary two-dimensional picture in such a way that a comprehensible message emerges. If we look at the pictograms in Figure 32 we immediately know intuitively when a picture is meaningful and when it is not; this is hardly surprising, since we are dealing with pictures that were made for man by man! But for automated picture decoding of any given text, which may contain a picture, this intuition must be replaced by a formal criterion.

Soviet researchers at the Institute for Linguistics in Moscow have in recent years developed just such mathematical procedures for the decoding by computer of picture-like messages. Methods have also been developed by military intelligence to help trace vital information in their immense files of photographic data as quickly as possible: the computers eliminate insignificant portions of the picture and emphasize edges and lines, which may reveal tanks or rockets. Today the technique of picture evaluation has advanced to such a point that practically any chosen structure can be programmed for identification by automatic processes. For example, cancer detection in medicine: electronic analysis programs can examine photographs of sections of tissue prepared by histologists for specific deformed cells which may indicate cancer.

But the examination of discrete series of symbols—that is, pulsed signals—for an unknown resolution into the length and width of a picture is beyond the power even of modern computers. The Soviet procedure for arranging the data in picture form is to select those pictures for which, after arranging the rows and columns, the length and breadth, in the 'correct' way, the greatest number of straight lines appear. These may take up any position in the picture. In complicated pictures the computer program then searches correspondingly for relationships of any kind between the lines.

The Arecibo cosmogram of 16 November 1974

Despite all the objections to cosmograms of the Drake type—most of which would certainly be known to him—a pictorial telegram of this nature, longer than the first suggestions and amplified with more biological information, was indeed sent into space.[24] The transmitter was the 305-metre radio telescope in Arecibo (see Figure 36); and the occasion the re-dedication of the telescope on 16 November 1974, after a 3-year programme of technical upgrading, which resulted in a marked increase in its sensitivity. Before reconstruction the telescope had a range of 6000 light-years: this is the distance over which the telescope was then capable of making contact with its counterpart. The adaptations increased the range by a factor of 10—suddenly the whole Galaxy was within range (at least in the band of the radio spectrum) for contact between two similar Arecibo telescopes aware of each other's positions. But a series of pulses lasting 3 minutes can at best be regarded as a symbolic attempt to set up interstellar communication. The (possible) impact of such efforts, as so often in dealings with 'extraterrestrials', is principally felt here on the Earth: in this case the initiative at least called attention to the new possibilities for a potentially imminent interstellar contact. The 1679 pulses were sent at a wavelength of 12.6 centimetres (a frequency of 2380 megahertz) with a signal rate of 10 bits per second, in the direction of a globular star cluster of our own Galaxy in the constellation

Figure 36 At present the largest on Earth, the Arecibo radio telescope fills an entire valley in the mountains of Puerto Rico with a reflector 304 metres in diameter. The primary focus assembly hangs 150 metres above the dish.

Hercules known as M13 (Figure 37). The reason for choosing this target—M13 is 24 000 light-years distant—actually had not so much to do with M13 as with the angular size of the radio beam sent out from Arecibo. The beam simultaneously covers all the 300 000 stars in the cluster—a most economical procedure enabling the project (in 24 000 years' time) to address all these stars at one blow. It is surprising that globular clusters had not previously been discussed as targets in the search for extraterrestrial contacts.

Carl Sagan, always an optimist, estimates that there is a 50 per cent probability of a civilization existing in M13 which is in a position to receive the Arecibo cosmogram. But the astronomers there 24 000 years from now would have to have aimed their telescope for 3 minutes, at exactly the right moment, at the present position of the Earth. The science journalist Günter Paul commented on this extremely unlikely possibility with the words: 'Under these circumstances it really does not matter whether the message is comprehensible at all.'[25]

The key to the Arecibo pictogram is the same as that for Frank Drake's first

Figure 37 The globular star cluster M13, 24 000 light-years from Earth. In 1974 a 3-minute message was sent to this cluster from the Arecibo radio telescope.

pictograms. The binary grid yields a comprehensible pictorial interpretation if we convert the series of symbols into a rectangle with the dimensions of the prime numbers 23 and 73 ($= 1679$). The numbers 1 to 10 in binary code run right to left from the upper right-hand corner (see Figure 38). This introductory strip is in part a brief mathematical primer, drawing attention to our decimal system, and has also the function of a call signal: the first data sequence unambiguously identifies what follows as an artificial message. The second row of symbols gives another indication: 1, 6, 7, 8 and 15. These are the atomic numbers of the elements hydrogen, carbon, nitrogen, oxygen and phosphorus, the most important chemicals for life on Earth. But here, at least, a change has been introduced, as compared with Drake's first cosmograms: the anthropomorphic and hence deceptive pictorial representations of Bohr's atomic model have been omitted in favour of the more abstract—and therefore perhaps more easily interpreted—atomic numbers.

This information is used in the block of symbols below to describe the

```
00000010101010000000000000101000001010
00000010010001000100000100101101010101
01010101010100100100000000000000000000
0000000000000001110000000000000000000000
1101000000000000000001100111001000000000
0000000101010000000000000000000111110
0000000000001000000000000000000000110000
1110010100000110000000000000000000001010
0001100001100011000110010111101111111
0111101011111000000000000000000000000000
0100000000000000000000000000111111000
0000000000011110000000000000000000000
001100000110001110001100010000001000
0000000011011011001100011110011011011
110111101111011011111110000000110011111
00000000010000001100000000000110000000
0011100000000100000001000001101000010000
1111000000000001100100011111010000000000
0000000000000100000000100000000000000001
0000001100000001000000001100001000000000
100000000000001000100000000000001000000
0110010000000000011000000010000010000
000011000011000100000000010000010001
0000001100000010100000000000001000000
1000010000000001110000000010000010000
0000011100000000110000000011000000000
01000111010101011100000000110011001101
01011100010110110000000100111001001111
1110111100000011000001101111000000010
1000000111101100001000010001001010001000
00000000000000000000000000000000000010
00010000000000001110100010101010101
010011000000000010101010000000000000
0101000000000010011110000000000000000
0001111111100000000001110000000111
000000000110000000011000000011010100
000000010110000011001110000010100011
0010000101000000100011000001101010001
00100001000000001010101000000000010000
010000000000010001000000000110001000
11101001111000
```

The 1679 bits of the
Arecibo message of
16 November 1974

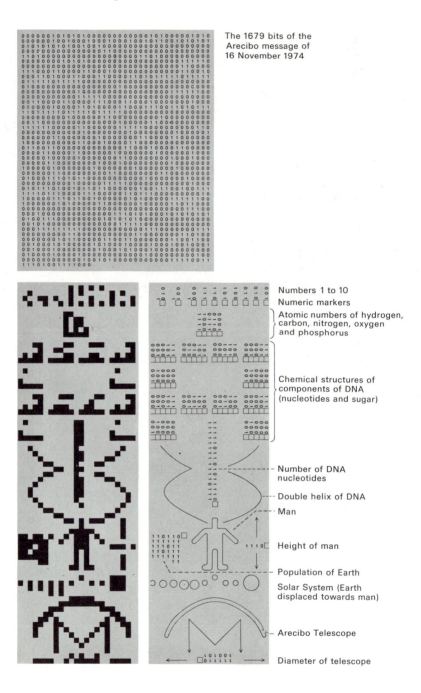

Numbers 1 to 10
Numeric markers

Atomic numbers of hydrogen,
carbon, nitrogen, oxygen
and phosphorus

Chemical structures of
components of DNA
(nucleotides and sugar)

Number of DNA
nucleotides

Double helix of DNA

Man

Height of man

Population of Earth

Solar System (Earth
displaced towards man)

Arecibo Telescope

Diameter of telescope

Figure 38 The Arecibo message consisted of 1679 binary pulses. It was transmitted on 16 November 1974 in the direction of the globular cluster M13, which consists of more than 300 000 stars. This astronomical object was chosen because the angular size of the beam from the telescope just covers the whole cluster. The '1' and '0' symbols were transmitted as 'on/off' signals, similar to Morse code. Arranged as a pictogram, they describe the biochemistry of man and other information concerning mankind and the Solar System (see text for details).

molecular components of DNA (deoxyribonucleic acid) (see Part I: 'The primeval soup and evolutionary reactors', page 38). Rows 32 to 46 show in pictorial form the double-helix structure of DNA, consisting of around 4000 million components. This number is given in the exact centre of the double helix. The most important structures of organic life on Earth have now been given a symbolic representation. A schematic figure of a man is placed beneath the double helix, with an indication of his size given by the number 14 (meaning 14 times the transmission wavelength), and to the left of the figure is an approximate value for the human population—about 4000 million. Beneath this is a diagram of the Solar System. The third planet is given special prominence by being positioned out of line—directly beneath the human figure. A downward-pointing cross-section of the Arecibo telescope and an indication of its size fill the remainder of the cosmogram.

Whether this mixture of pictures of man, Solar System and telescope, together with abstract numerical material, constitutes a code which lends itself particularly well to decipherment, is no longer easy to decide. The experts in cosmic linguistics seem to have become too familiar with this kind of picture puzzle in the meantime to remain unprejudiced about it.

The search for contact as a two-party game strategy

In choosing the most promising from among many possible ways of sending a message and of searching for artificial signals, we cannot avoid making certain assumptions about the sender and the receiver. An optimum *strategy* for the search programme of the receiver can be devised according to the nature of the transmission thought probable. Unfortunately there is no optimum strategy to cover all possibilities short of a TOTAL programme: observing and transmitting continuously in all directions and on all frequencies. This situation has much in common with the choice of an optimum strategy for winning games, for example a game of chess. A special branch of mathematics called game theory is concerned among other things with the optimum winning (or losing) strategies for a given strategy of the opponent. The 'opponent' facing us in our search for contact with a communicative extraterrestrial civilization is the technology of the sender and his transmitting programme; the 'protagonist' is the structure of the receiver and the way he conducts the search; the 'winnings' which are played for, and which must be guaranteed as far as possible by the choice of optimal moves, can be regarded as successful proof that a technically advanced civilization exists somewhere outside our Solar System, in other words the successful establishment of contact. In addition to simple evidence of an extraterrestrial civilization, the aim is also, of course, to learn as much as possible about this civilization, to receive signals from it, to decode them, and perhaps also to send back deliberate messages about ourselves, inhabitants of the planet Earth.

A TOTAL search and sending programme is at present not possible. Practically all astronomers and astrophysicists would have to devote themselves solely to this task; all available telescopes would have to watch the entire sky exclusively for artificial signals—and between times to beam out suitable messages; many thousands more giant radio telescopes would have to be built to make sure that the programme could continue 24 hours a day—a whole nation, perhaps all mankind, would have to devote itself almost exclusively to this task. Forced to limit our investment of money and personnel, but nonetheless demanding that the greatest possible chances of success be assured, we must devise an optimal, consistent strategy.

On top of this, the would-be receiver can, of course, only guess at the strategy of the sender. The search for interstellar signals is more like a blindfold game of chess, in which we do not even know by what rules our opponent is playing—indeed, not even whether he is playing at all!

The problem, then, is to choose from the multitude of possibilities those which hold out the greatest promise of success.

But if the conceivable possibilities exceed our operating capacity, the range of these possibilities must be reduced by making assumptions, by criteria of simplicity and plausibility. The assumptions may be questionable, but with their aid we can at least develop feasible projects. This principle of methodical strategy is a standard procedure in theoretical and experimental science.

Cosmic linguistics, therefore, might start from the assumption that galactic signals from other civilizations which we are intended to receive and translate have not been *deliberately* made hard to understand. We start, therefore, from the following simple assumption about 'non-decoding'.

Principle of anti-cryptography

A (purposely sent) interstellar message will be so designed in its structure and in the nature of its transmission as to maximize the probability of its discovery, whether by deliberate searches or through accidental observation by astronomers.

In 1961, Sebastian von Hoerner drew attention to another plausible criterion of simplicity.[5] He correctly thought other civilizations equally unable to squander arbitrary amounts of energy for their activities. Even supercivilizations must economize; nowhere in the Universe is unlimited energy available, and the law of the conservation of energy is just as valid in the farthest corner of the Universe. The cheapest thing, of course, is total passivity: interstellar silence. The following formulation of this idea seems to me to be appropriate:

Principle of economy

Civilizations will carry out such activities as they may for any reason wish to undertake, and for which they have the technology, by the most economical means.

Other civilizations, too, will develop methods which make use of the least personnel (if there are persons), material and energy. This is especially true for societies living on planets. Planets have limited surfaces and thus limited supplies of food and energy. And other civilizations, spurred on by competition for these supplies, will have undergone evolution. This universal evolution therefore not only results in a dominant intelligent species, but implants in it at the same time a tendency to save, the principle of economy.

Another assumption is borrowed from cosmology:

Principle of normality

The existence of our civilization, including its present technology, represents a typical phenomenon in the Universe.

This principle is a version adapted for biology and civilization of the Copernican principle, first published by Copernicus in 1560, shortly before his death. Essentially it states that the Earth is not at the centre of the Universe. The Copernican scheme supplanted the geocentric view of the world and banished the Earth to an unimportant outpost of cosmic events—according to Freud, one of the three fundamental humblings of mankind (beside Darwin and Freud himself). So long as we know no better, this principle also proposes an analogous normality for the origin of life on Earth and the development of the human race, including its technology. Since we are a young civilization, it directly follows that many other civilizations must be in possession of far more advanced technology and find things technically simple which still seem difficult to us. As will be discussed later, I consider the hypothesis formulated in this principle to be the most questionable of all the simplifying assumptions. It seems, indeed, at least so far as technically advanced societies are concerned, even to contradict the available observations.

It is an immediate consequence of this hypothesis—and the following discussion will be based upon it—that the present state of Earthly technology cannot be a measure of the strategy of interstellar communication methods. Rather, we must use as criteria the extremes of possibility, the limitations imposed only by the laws of physics and the nature of the Universe, to which all civilizations are subject.

The principle of economy also has far-reaching conceptual consequences. A rocket which is to fly to the nearest star, Alpha Centauri, at 99 per cent of the speed of light, land and return again, must weigh about 100 000 million (10^{11}) tonnes if it carries its propellant with it from the first (see Part III). It costs only a few dollars, on the other hand, to send a 60-word telegram with a modest radio telescope to a similar telescope on Alpha Centauri. It is thus considerably cheaper to send information rather than material objects. Also, information is transmitted by electromagnetic waves with the greatest possible velocity, the speed of light. Usually, therefore, it will be more economical to send the plans for the construction of objects than the objects themselves.

One aspect of this principle—and this is the idea behind the strategy—is that the sender will try to reduce all unknown parameters characterizing the signal to a minimum. What are the unknown parameters to be established by the astronomer? They are these:

1. Distance.
2. Direction.
3. Time: duration and number of repetitions.
4. Frequency, bandwidth, signal rate, polarization of the waves and their modulation (code).[26]

Strategies for distance, direction and frequency

Interstellar search programmes so far undertaken by radioastronomers started from the direct investigation of stars like the Sun in our immediate galactic neighbourhood (see 'Project Ozma and its successors', page 155). As the performance (sensitivity) of the radio telescopes increased, their range expanded (see 'Projects of today: Cyclops', page 162), but the strategy—to listen only to individual stars—has remained the same. This proceeding is in tune with human nature: an idea is tested on a small-scale model, the experience thus gained and knowledge won in the meantime are applied to a project of the next order of magnitude, and so on up to the level allowed by contemporary technology (or financial resources).

The opposite strategy is to listen in to the whole sky at the same time. This is done with 'isotropic' aerials, which can receive radiation simultaneously from all directions. The advantage of an isotropic search programme is that the number of stars which can be reached increases in an optimal relationship to the distance. Up to about 1000 light-years away the number of stars increases as the cube of the distance, but more slowly further out. The reason is that the Sun lies almost on the edge of the spiral of the Galaxy, almost exactly on the equatorial plane of the galactic disc (seen in profile) and this disc is only about 4000 light-years thick.

But signals become harder to detect with increasing distance, as the intensity of the signals decreases, and this must be compensated by making greater demands on transmitting efficiency and the sensitivity of the receiver. And beyond a few hundred light-years we must rely on the statistical values of the frequency of Sun-like stars with habitable planets. Finally, the chances of ever establishing two-way communication diminish rapidly with increasing distance.

According to Nikolai Kardashev at the Institute for Cosmic Research in Moscow, the following statements regarding civilizations provide the basis for an optimum strategy for distance:[6]

Capitalist perspective

All civilizations reach a stable state below our present technological level, or destroy themselves and die out.

Evolutionary perspective

Civilizations develop (technologically) beyond our present stage if they first succeed in overcoming all phases of potential self-destruction, and secondly, if they succeed in developing a daughter civilization outside of their Earth, which produces a surplus.

The first perspective is derived from the development of technology in western civilization (hence the adjective 'capitalist') and its future prospects—currently most uncertain, at least in part—of growth and continuously increasing energy consumption. It is assumed that nuclear fusion is unsuccessful, fast-breeder reactors are too dangerous as uranium-saving successors to pressurized water reactors, and totally new sources of energy on a large scale are not to be found. In that case it is very questionable whether a leap to a supercivilization is technically and politically possible *for us* at all.

The second perspective, on the other hand, assumes that, after the technological crisis which will confront us within the next 100 years, the world will undergo further rapid development, rather on the lines of Freeman J. Dyson's visions of the future (see 'The astroengineers of supercivilizations', page 200).

The following strategies, corresponding to these two perspectives, now offer themselves:

Distance strategy A

Search for civilizations comparable to our own (Kardashev's Type I, see page 90), situated in a planetary system and provided with transmitters of terrestrial capability. Search for signals of small bandwidth (monochromatic) from stars like the Sun. Investigation of nearest stars first, then selected stars at greater distances.

Distance strategy B

Search for supercivilizations (Kardashev's Types II and III, see page 90) equipped with more powerful transmitters and in a position to manipulate planetary systems and stars (astrotechnology). Initial search for the most powerful and often also the most distant sources of radiation in the Universe. Only when it has been established that they are merely natural sources should fainter objects also be investigated. Parallel search for narrow and broadband artificial pulsed signals from the centre of the Galaxy and the neighbouring galaxies and quasars; search for signs of astrotechnology; search for new types of objects in ranges of wavelengths still little explored (such as the infrared and ultraviolet), and investigation of individual, unusual and abnormal astronomical objects.

Strategy A has the advantage that we can be sure of discovering the civilizations nearest to us in space by these means—and it is only with our very closest 'neighbours' that we have any chance of accomplishing two-way communication, at least within the astronomically 'short' time of a few centuries. Project Ozma (see page 155) was in fact built on just this strategy. Besides this, the current state of radioastronomy is so far advanced that now, in a sense, the whole of the Galaxy lies 'open' to us.

In this, an amazingly rapid development lies behind us (see Figure 39). In 1945 a radar signal was reflected from the Moon for the first time; in 1959 the planet Venus was first located by radar. And two Arecibo radio telescopes could today make contact with one another all the way across the Galaxy.

If both strategies seem to hold out equal promise, they should both be put to work independently of one another. All the projects succeeding Ozma, up to the search programme initiated in 1978 at the Ames Research Center (see 'Project Ozma and its successors', page 155), followed Strategy A, whereas the survey of four neighbouring galaxies by Drake and Sagan in 1975–76 adopted Strategy B. If supercivilizations of Type II or III are converting large amounts of their energy into radio waves, these civilizations should in principle have been detected by Drake and Sagan.

For the present, we cannot expect mankind to pave the Earth and its surroundings (the Moon) with the thousands of Arecibo radio telescopes which would be needed to keep an eye on the whole sky fairly completely and systematically.

It is only through further simplifying criteria that a practical programme can be derived from these global strategies. In the opinion of radioastronomers, the *transmission* of interstellar messages should satisfy the following conditions:

1. The energy used for the transmission of one unit of information (one bit) must be kept to a minimum.

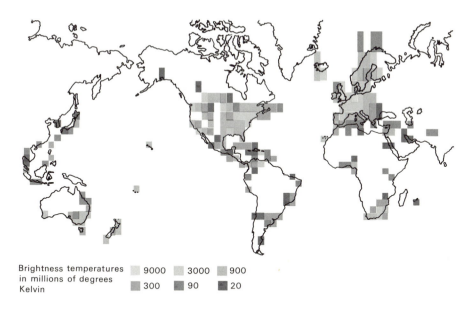

Brightness temperatures
in millions of degrees
Kelvin

9000   3000   900

300   90   20

Figure 39 In the radio region the Earth has for some years been brighter than the Sun. Radio and television stations transmit between the frequencies of 40 and 220 megahertz. The intensity of this radiation is shown on the map averaged over small squares measuring 5 degrees of latitude by 5 degrees of longitude. The resulting radio brightness can be represented as the temperature to which the corresponding portion of the Earth's surface would have to be heated to emit this radiation thermally. (Adapted from 'The Search for Extraterrestrial Intelligence' by Carl Sagan and Frank Drake. Copyright © (1975) by Scientific American, Inc. All rights reserved.)

2. Interference from other sources in the neighbourhood of the transmitter must be as weak as possible.

3. The ratio of signal to 'noise' (the natural radiation present at the transmission frequency) must be greater than one.

4. The duration of the transmission must be as short as possible.

A *call signal* transmitted before the actual message (see 'Artificial message or natural noise?', page 99) must also make it as easy as possible for any interested listeners within the Galaxy to filter out the whole message from the background noise and then to decode it.

Besides the strategies for distance detailed above, then, we must also rely on other assumptions concerning the transmission methods of a galactic civilization. The assumptions may be wrong, but each at least indicates a clear course of action. It then depends on the choice suggested by the scientific, cultural and political spirit of the age, which procedure is chosen as the 'most reasonable'.

Directional strategy A

Assumption: *both transmitter and receiver are operating with isotropic (non-directional) aerials.*

In this situation, which is unfavourable to long-distance transmission, there is no preferred relationship between transmitter and receiver. The transmitter radiates whatever he wishes to communicate uniformly in all directions, the receiver listens in all quarters of the celestial sphere. This is, so to speak, the poor man's version of TOTAL strategy; we must be content with a small range for the search.

Directional strategy B

Assumption: *only the transmitter radiates isotropically; as receivers we select a direction that seems promising to us.*

The receiver limits himself to receiving beamed information from a small section of the sky defined by its celestial co-ordinates. This merely requires that we can construct aerials with well-defined directional properties.

Directional strategy C

Conversely to strategy B, it is here assumed that *the transmitter sends directional signals, but the receiver anticipates signals from an unknown direction.* Therefore the receiver, Earth, receiving signals equally in all directions, will survey the whole sky with isotropic aerials.

Directional strategy D

Assumption: *transmitter and receiver know their respective co-ordinates and direct their aerials at one another.*

Each of these four strategies leads to an optimum procedure for transmission and receiving (if the radio noise is known for transmitter and receiver). While strategy A assumes no previous knowledge of the direction in which an advanced civilization is to be expected, the other directional strategies presuppose an increasing amount of such knowledge. The direction of a neighbouring civilization is probably the most important information in a search for contact; ignorance of the direction of a civilization capable of contact impairs the process of searching and sending more than any other factor. Although a continuous search in all directions (directional strategies A and C) has the advantage that at least no part of the sky will be left out, the advantage of greater range must be given up. But if we are driven by limitations of observing time to select directions, the regions of the sky which contain the most stars recommend themselves: the equatorial plane of the galactic disc and in particular the galactic centre itself.

Within 1000 light-years of the Sun, the number of Sun-like star candidates in a direction perpendicular to the Milky Way is about 70 per cent of the number in the direction of the galactic equator.

For directional transmission and reception (directional strategies B, C and D) considerably stronger signals can be sent in the desired directions, and thus can be transmitted to nearby stars in an optimum manner. On the other hand, if a civilization sends a signal to Earth which arrives just when our telescopes happen to be searching in some other direction, we will simply miss the signal, irrespective of how strong it is. The situation is like that of two needles in a haystack looking for each other. So long as both needles are searching at the same time, they are hardly likely ever to find one another. But if one needle sticks firmly in one position, and the other needle sets off on a systematic search, this strategy will inevitably, in the end, produce success. But which civilizations are keeping still, waiting to be discovered, and which are committing themselves to active search? As this cannot be agreed in advance, any two galactic 'needles in a haystack' which hope to find each other must simply do both at once: transmit (sit still and call attention to themselves) and receive (search).

Strategies for transmission time, bandwidth and signal rate

The number of conceivable different methods of transmission is almost overwhelming: continuous transmission or periodic signals, which are thus regularly repeated, or once-only transmissions. So long as the signal is transmitted in all directions—isotropically—the message will reach every potential 'subscriber' for as long as the sender transmits it. For beamed radiation with marked directional characteristics, there arises the question of how long transmission should continue in particular directions. For radio telescopes like the parabolic reflector at Arecibo, which are firmly fixed in the ground, the beamed nature of the transmission means that during the 24-hour rotation of the Earth a given point in the sky is only covered for an interval of a few seconds. Only if the telescope is partially steerable can one point in the sky be followed for somewhat longer.

Communication over long distances is impossible at frequencies below 1 megahertz, due to interstellar absorption; the optimum region is between 1 and 10 gigahertz. In principle, the radio noise in the atmosphere of our planet can be eliminated by building telescopes in Earth orbit or on the Moon.

The detection of a message may also depend on its bandwidth, that is, the frequency interval over which the energy of the transmitter is distributed. The broader the frequency band we choose as telescope observers, however, the greater is the danger that a narrow-band signal will be overlooked: if it reached the Earth in a very weak state, it would simply be swamped. In their search for artificial interstellar singals, radioastronomers start from the premiss that such signals will be distinguished by very narrow frequency bands. There are at least two considerations to support this: first, most natural wave sources radiate in broad bands, and secondly, a signal of given transmission energy is more easily detected, the narrower the band with which it is transmitted, if only it is picked up by a receiver with a suitable frequency band.

The search for narrow-band signals has one great disadvantage. The narrower the frequency band of our receiver, then clearly the more time is occupied in combing through an interesting range of frequencies, such as, for example, the microwave window between 1 and 10 gigahertz. For signals with a bandwidth of 1 hertz (which is already broader than for most terrestrial radio and television transmitters) there can be 10 000 million (10^{10}) such signals side by side in the microwave window. If we could examine a band 1 hertz wide every second, to go through only once would take nearly 300 years!

If we tune the electronics of our aerial to narrow-band signals, we must be prepared for another effect. If the transmitter is installed on a planet or a satellite, as would be expected for a Type I civilization, the frequency of the signal will be periodically Doppler-shifted by the rotation of the planet and its orbital motion around its star (see Figure 40). The frequency may be shifted by up to several hundred kilohertz, at a rate of several hertz per second. Such frequency modulations, however, would help to distinguish the signal from cosmic radiation and Earth-bound interference.

To escape from this dilemma, either well-defined 'magic' frequencies are chosen (see below), as with Project Ozma, or a technical wonder known as the multi-channel spectrum analyser is used. With this instrument, a large number of adjacent bands ('channels') can be measured simultaneously. The analysers developed for the very latest search programme of the Jet Propulsion Laboratory and the Ames Research Center are capable of recording data simultaneously on a million channels each of 10 hertz bandwidth, or alternatively on 1000 million channels each of 0.01 hertz bandwidth (see 'Project Ozma and its successors', page 155).

For individual observations the analyser is operated with 1000 million channels at a bandwidth of 0.3 hertz. Observations are thus made possible at one blow over the complete 'waterhole'—the stretch of frequencies between hydrogen (H) and the hydroxyl (OH) radical. This system has only been made possible by the rapid development of micro-electronics. 'In many respects the whole program has only become possible as a result of the current explosion in the field of micro-electronics. Today we are able to accommodate in an electronic frame in the space of a few centimetres what would have filled whole buildings with electronics not so long ago,' says Robert E. Edelson, leader of the search project at the Jet Propulsion Laboratory.

The projects of the two American research groups—the one a survey of the

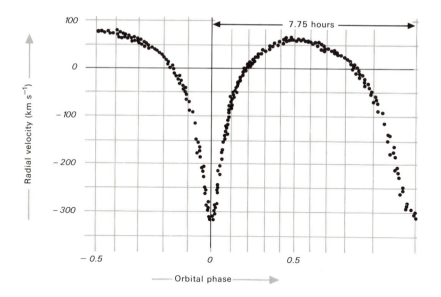

Figure 40 Signals from pulsar PSR 1913 + 16 as it orbits its binary partner. The dots represent measurements of the pulse period. When the pulsar moves towards us, the arrival times of the individual pulses draw closer together, owing to the Doppler effect; when it moves away from us, the pulses draw further apart. The pulse interval is thus a measure of the radial velocity of the pulsar. The interval between minima yields the orbital period. The deviation of the curve from a sine wave shows how strongly elliptical the orbit is. The discoverers, R. A. Hulse and J. H. Taylor, calculated an orbital period of 7.75 hours and a value for the eccentricity of 0.61.

whole heavens, the other the study of single objects—are the most thoroughgoing current search programmes for artificial signals from extraterrestrial intelligences. (They started at the end of 1978 and will run for about 5 years.) With only a few per cent of the observing time hitherto devoted to initiatives of this type, the new projects already exceed in range all previous investigations.

At the same time the scientists took into account the possibility that the extraterrestrial civilization might also be sending its message from the surface of a planet and thus through its atmosphere. At frequencies less than 1 megahertz or greater than about 30 gigahertz, all electromagnetic waves are absorbed by typical planetary atmospheres.

But there are plenty of other disturbances within this frequency range with which not only terrestrial radioastronomers but also all other civilizations have to live: on the low-frequency side there is the 'radio noise' of the Galaxy, caused by the radio radiation of all its stars; on the high-frequency side there occur the radio lines of molecules in the atmospheres and the interstellar gas clouds, as well as the 'quantum noise' of high-energy electromagnetic radiation, and finally the noise induced by the transmitter and receiver.

The radio spectrum of the sky as observed from the Earth's surface is rather 'noisy'. Every radio telescope receives the 3 degrees Kelvin background radiation, the relic of the hot early stage of the Cosmos at the time of the Big Bang. This radiation falls off slowly at about 60 gigahertz, and here the 'quantum noise' of the photons, associated with all electromagnetic radiation, begins to dominate. The radio noise of the Galaxy principally arises from synchrotron radiation by charged particles (electrons, for example) spiralling about the lines of force of magnetic fields. All these radio sources leave a relatively quiet zone in the frequencies between 1 and 100 gigahertz, which will appear almost identical to all observers near stars like the Sun. The Earth's atmosphere also emits radio waves, as water and oxygen molecules absorb and emit energy at 22 and 60 gigahertz respectively. All these radio sources added together produce the heavy line in Figure 41.

But all this unwanted noise still leaves a large gap, a wide minimum, in which only the characteristic radiations of certain atoms and molecules abundant in the interstellar medium intrude. The article in *Nature* by the Cornell physicists Cocconi and Morrison[3] contained a fascinating proposal for a frequency within this range which might be particularly suitable for interstellar communication: that of the radio waves emitted by atomic hydrogen at a wavelength of 21 centimetres, which had been discovered 8 years before in the radio noise of the interstellar medium.[27] This radiation has its origin in the hydrogen atom itself, which consists of one electron and one proton. As the electron orbits around the proton its so-called 'spin' can possess one of two states, and is constantly jumping back and forth. The minute energy difference as it 'flips over' corresponds exactly to the wavelength of 21 centimetres.

Hydrogen is the most abundant element in the Universe, and accordingly the hydrogen line is the strongest resonance line. Astrophysically, too, this line is important: by its distribution, hydrogen gives more detailed information about the structure of the Galaxy than other radiating molecules. This physical fact is true everywhere in the part of the Universe known to us—so it can be recognized by any other civilization preparing to send signals into space. Thus, the centimetre range of radio waves should appear especially suited for interstellar communication to most galactic astronomers. Near the hydrogen line (at 1.42 gigahertz) the hydroxyl radical radiates, at 1.665 gigahertz. The range of frequencies between these two lines is the so-called 'waterhole'.

Although the hydrogen line seems to offer an optimum frequency for the transmission of interstellar messages, closer examination shows that it is by no means so useful. Arguments against its use have been presented by the American radioastronomers Dixon[26] and Drake.[22] For one thing, the very intensity of the line could conceal artificial signals. Conversely, the transmission of artificial

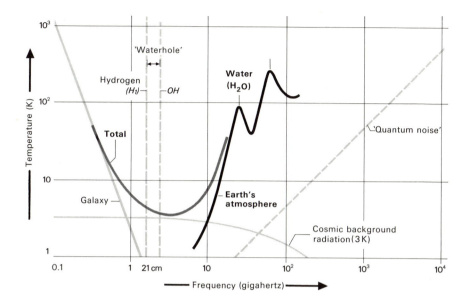

Figure 41 Atmospheric and galactic radio noise leave a gap suited for communication. The radio spectrum of the sky, as measured from the surface of the Earth, is rather 'noisy'. All radio telescopes receive the 3 degrees Kelvin (K) background radiation, the residuum of the hot early stage of the Universe in the Big Bang. This radiation falls off slowly at about 50 gigahertz, where the 'quantum noise' of photons, associated with all electromagnetic radiation, begins to dominate. The radio noise of our own Galaxy principally arises from synchrotron radiation by charged particles (for example, electrons) spiralling around the lines of force of magnetic fields. All these radio sources leave a relatively quiet zone in the radio spectrum in the region between 1 and 100 gigahertz: the hydrogen line occurs almost at the minimum of this quiet radio region. The wavelengths in the neighbourhood of the hydrogen line are seen by many radioastronomers as especially suited for the purpose of communication within the Galaxy.

signals at 21 centimetres will produce interference hampering astronomers on the same planet in their observations of the hydrogen radiation. For this very reason, frequencies near 1.42 gigahertz are already taboo for radio stations on Earth.

However, perhaps the line can still be 'rescued', if the search for civilizations moves to closely neighbouring frequencies. Civilizations may decide to develop only one of two activities: 21-centimetre astronomy or interstellar communication: this will depend on their respective interests and state of knowledge. Astronomers on Earth today would probably put the investigation of the hydrogen line in first place: other civilizations might make a different choice.

The other end of the 'waterhole', the OH emission, has also fallen into disfavour. In fact, it is divided into several closely grouped lines at 1.612, 1.665, 1.667 and 1.720 gigahertz (the so-called lambda doubling and the hyperfine structure of the lowest energy state of OH). Which line should we choose? This question brings a new unknown into our strategy concept. In short: the magic lines have lost their glamour.

Drake is investigating the factors governing the sensitivity threshold of a radio aerial: the radiation strength and tightness of beam of the transmitting aerial as well as the receiving aerial, and the distance from the source. Drawing on natural laws applicable to these factors, Drake shows that a transmission of a power which can just be detected by the receiver indicates a fairly definite frequency; and this *independently* of the particular level of terrestrial technology. Governed by the intensity of cosmic noise in different directions in the sky, Drake's elegant reflections give an interval of wavelengths between 3 and 8 centimetres, the optimum from this standpoint.

For each of these optimum wavelengths we can see furthest in the direction concerned. At the same time, this is the wavelength on which information can be transmitted most cheaply—Drake, in fact, started out from this requirement. We may now finally combine this conclusion with the philosophy of anti-cryptography—make the most obvious choice possible—and look for a spectral line among these optimum frequencies.

At one time the water line at 2.22 gigahertz was suggested. It is true that the disturbing background of atmospheric absorption is stronger at this rather higher frequency, but this would not effect telescopes outside the Earth's atmosphere. Besides this, higher frequencies (shorter wavelengths) bring with them the advantage of greater range.

Recently two American astronomers, T. B. H. Kuiper and M. Morris, have suggested defining a single fundamental frequency on a quite different basis from that of using particular molecular lines.[28] This frequency is to be arrived at by a suitable combination of the natural physical constants, which must be known to every civilization practising radioastronomy, and thus also physics. These constants are above all the velocity of light, the Planck constant and the fundamental properties of the electron. Kuiper and Morris recommend the special frequency of 2.55 gigahertz, corresponding to 1.2 centimetres.

Stimulating though the speculation of Kuiper and Morris may be on the problem of a frequency marked out 'by Nature', their special suggestion does not survive strict criticism. Too often physicists are influenced by their special fields and the associated theories. Other frequencies could be 'marked out' in various fields of physics, using arguments akin to those of Kuiper and Morris. Moreover, the famous fundamental constants of physics have had checkered careers.[29]

In the history of physics during the past 300 years the numerical values of the physical constants have been steadily improved, as advances in measuring methods led to refined or altered figures. And what is regarded by physicists as 'fundamental' is strongly dependent on the scientific spirit of the age. In the eighteenth century the velocity of sound was considered fundamental. We know today that the velocity of sound in a substance can be calculated from other fundamental properties of the material (such as pressure and density). Another striking example: scientists still thought in Galileo's time, that is, at the beginning of the seventeenth century, that light propagated at infinite speed. Only in the year 1675, after accurate observations of the moons of Jupiter, did the Danish astronomer Olof Rømer conclude that the velocity of light must have a finite value.

The significance and numerical value of the so-called constants of nature thus depend clearly on the historical period in which terrestrial scientists are concerned with nature. And it would be a mistake to assume that we have come to the end of this process. Certainly our knowledge of the constants of nature involves no timeless values. Thus any fundamental frequencies deduced from such constants of nature are correspondingly questionable. The alternative method, observing numerous adjacent frequency bands with a multi-channel system, seems to me less problematical and more promising. This latter technique is in fact being used in the important, highly endowed large-scale search by the Jet Propulsion Laboratory and the Ames Research Center.

In the case of unintentionally transmitted signals, it is more difficult to find a frequency strategy which will be of some use to us. And this is probably the more realistic side of the situation; the Earth, too, has hitherto sent out no deliberate call signals. We must therefore allow for the possibility that *nobody* is deliberately transmitting, and *everybody* is only listening. In such a situation there is no optimum frequency. Moreover the range of the signals will be small, since for practical purposes the transmission is non-directional.

Evidence of chance radio signals from civilizations at a distance of up to 1000 light-years and more is hoped for. If, on the average, civilizations are statistically uniform in distribution, the number of civilizations that can be reached increases as the cube of the distance (the additional volume of space). (On account of the flattened form of the Galaxy this holds good only within a few thousand light-years' radius.) But here there are two problems: first, to discover artificial signals, and then to extract meaningful messages from them. We have seen that artificial signals can be detected over distances at which the *message* can no longer be filtered out from the noise. The exchange of cosmic messages is accordingly limited on the one hand by the range of simple signal detection, on the other, by that for meaningful communication.

No communication at all is possible, of course, until a signal has been detected. To L. M. Gindilis the communication range is 'the maximum distance over which the communication system is capable of sending or receiving information with a given degree of reliability (or confidence level)'.[6] Beyond this distance the signal can no longer be reliably reconstructed.

The communication range is also greater, the narrower is the frequency band of the signals, and naturally also the stronger the transmission, the tighter the beam, and the more sensitive the receiving equipment. This will in fact be at an *optimum* if sender and receiver tune their frequency bands to one another, so as to work with matching frequency bandwidths. After an artificial signal has been detected, the receiver, in an ideal strategy, will match himself to the transmitter along these lines; he then extracts the transmitted information with the greatest certainty possible in theory. The maximum signal range will then exceed the maximum distance for communication by a factor of 10 to 100!

Given similar frequency bands in transmitter and receiver, the expenditure of energy is least when transmission and reception are carried out with directional aerials (for example, Arecibo). Following the economy principle, this is the optimum energy strategy. On the other hand the cost of this is that contact may not be established at all. Only when the two participants in the exchange are informed in advance of each other's position has this strategy any reasonable prospects of success. As already mentioned in the discussion of directional strategies, there is, besides, only a realistic chance of making contact if the transmitter, at least, sends isotropically. The receiver can then still use directional aerials to explore the whole sky. Bridging 1000 light-years will cost the transmitter a level of radiation power such as a Type I civilization has available (10^{13} to 10^{17} watts).

For pulsed signals, reception by directional aerials again presents problems. For transmission over large distances the sender must make the pulse duration so long that the information content falls to one bit per hour, or even one bit per year. If the directional aerial is only directed for a short time at a source of slowly pulsed radiation it is quite possible that the signal will simply be overlooked. This means that aerials without directional characteristics are better suited to the detection of pulses. The disadvantage of such aerials is their small range, which can only be compensated by increased transmission power. For a given distance of 1000 light-years the energy required of the transmitter rises to about 10^{15} to 10^{23} watts—unless the receiver decides on a larger investment and covers the sky with a battery of several thousand directional aerials.

Once the civilization has discovered signals from another race, it can at once turn to simple directional reception. If this discovery is made simultaneously in the opposite direction—which would be actual establishment of contact—information at the rate of 1 to 10^8 bits per second can be transmitted over a distance of 1000 light-years with from 10^6 to 10^{14} watts—a much lower strength of transmission than was needed for the initial detection of the signal.

'This apparently paradoxical consequence is quite easy to explain. The signal strength necessary for the detection of the signal is the price we have to pay for the address of the subscriber.'[6]

We now turn, however, from the discussion of strategy to the concrete investigations in which mankind has hitherto attempted to track down radio signals sent by extraterrestrial beings.

Project Ozma and its successors

The year after Cocconi and Morrison[3] drew attention to the possible value of the 21-centimetre hydrogen line for interstellar communications, recommending that the search be especially concentrated on the nearer stars outside the equatorial plane of the Galaxy, Frank Drake, the then Director of the National Radio Astronomy Observatory (NRAO) at Green Bank, published in the magazine *Sky and Telescope*[30] the first concrete plan to monitor stars for intelligent signals in the region of the 21-centimetre line.[31] He gave his project the name Ozma, after the queen of the imaginary land of Oz in the children's story by L. Frank Baum: a very distant land, difficult to reach and inhabited by exotic beings. As targets for his search programme Drake selected stars designated by NASA physicist Su-Shu Huang as most like the Sun among those near to us: Tau Ceti and Epsilon Eridani—one 11.9, the other 10.8 light-years away from Earth.[27]

Early in 1960, at 4 a.m. on 8 April, Drake put his programme into operation— by no means to the satisfaction of all scientists. In May, June and July—a true total observing time of about 4 weeks—Drake searched for artificial signals across a 400-kilohertz range of frequencies in the neighbourhood of the hydrogen line, with the narrow bandwidth of 100 hertz. During this period no artificial extraterrestrial signals were detected (Figure 42); thereafter the search pro- gramme was discontinued, as the telescope was required for other radioastro- nomical work. Drake himself was by no means surprised by the negative result: another civilization as such a close neighbour in the Cosmos would of itself have presented a case of galactic over-population. Nonetheless, he considered the expenditure justified for a pioneering project.

Project Ozma was dead, but the sceptical world of science was now less disposed to dismiss ideas and investigations in this field as products of science fiction or pseudo-scientific cosmological daydreams. For some years after this, however, no further systematic astronomical observations were undertaken which might have withstood strict scientific criticism.

In 1964 the Soviet radioastronomer Nikolai Kardashev, then at the Sternberg Observatory, called on his colleagues at the other radio telescopes to investigate the astronomical object designated CTA 102 for artificial radio signals. Scarcely a year later, in April 1965, Kardashev and two of his associates asserted in the Russian *Astronomicheskii Zhurnal* (*Soviet Astronomy*) that they had filtered out an artificial signal from the radio emissions of CTA 102. 'We have recorded radio oscillations of constant period which repeat themselves every hundred days. The data have been checked by our research team with the greatest care and leave no room for doubt,' wrote Kardashev's collaborator Gennady Sholnickii.

Epsilon Eridani

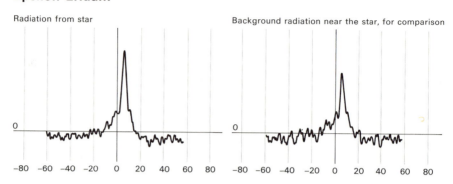

Figure 42 Project Ozma: a comparison of the radio emission from Epsilon Eridani with that from a random section of the background shows no differences which would indicate the presence of artificial signals.

Even before Project Ozma, however, American scientists had already thought that they had intercepted artificial signals from Jupiter and Venus. The supposed Venus signals were recorded 3 months before the first Sputnik was launched on 4 October 1957. The star Vega, 27 light-years from the Earth, was also regarded for a short time as the source of artificial messages. Two Harvard astronomers had recorded the radio noise from the direction of Vega for 26 months, and found that curious signals occurred every 4 days at the same wavelength. In their opinion these could only be of artificial origin.

Project Ozma itself was not safe from surprises of this kind, although every precaution was taken to eliminate all possible terrestrial influences by the best possible filtering of the data. Drake and Cocconi, for instance, also took part in the experiment; shortly after beginning their new assignment and after several hours' fruitless observation of the star Epsilon Eridani, they turned the telescope to Tau Ceti. Soon after starting the new run they registered a series of regular signals for several minutes—about eight pulses a second! But before the equipment had been precisely adjusted to examine this phenomenon more closely, the signals disappeared. When the occurrence was repeated a fortnight later, it was decided that the source of the portentous signal could only be a terrestrial transmitter—probably a secret military installation.

In all cases where the detection of artificial signals has been announced, the discoverers have eventually withdrawn the claim that they have intercepted intelligent radio signals from space—in most cases a terrestrial source or a natural phenomenon has been identified as the cause.

Sobered by these disappointments, scientific interest in the detection of

Table 7 Astronomical objects investigated for interstellar communication in Gorki in 1968*

Object	Distance in light-years	Spectral class	No. of observations
Epsilon Eridani	10.8	K2 V	66
Tau Ceti	11.9	G8 V	6
380 Ursae Majoris	14.7	dM0	2
Rho Comae Berenices	27.2	G0 V	7
Beta Canis Venaticorum	30.2	G0 V	7
Eta Boötis	31.9	G0 IV	6
Iota Persei	38.8	G0 V	4
47 Ursae Majoris	44.7	G0 V	10
Psi-5 Aurigae	48.66	G0 V	2
Pi' Ursae Majoris	50.2	G0 V	4
Eta Herculis	61.6	G8 III–IV	3
M31 (Andromeda Galaxy)	2.5×10^6	—	6

* From ref. 4.

intelligent signals from extraterrestrials seemed to wane. For several years nothing happened to re-kindle the researches begun by Project Ozma. Western radioastronomers were thus a little surprised, on attending the Byurakan Conference in the Soviet Union in September 1971, to hear the Soviet astronomer Vsevolod Troitskii announce the preliminary results of a Soviet Ozma project.[32] At Gorki University a 15-metre radio telescope was available to astronomers for their researches. They had added to this a receiver specially developed for the reception of radio waves of 21 and 30 centimetres. This receiver permitted the simultaneous recording of data on 25 different frequencies separated by 4 kilohertz, with a bandwidth in each case of only 13 hertz. Each observation had a total time of 10 minutes. Eleven Sun-like stars, up to a distance of 60 light-years away, were observed in this way on between 2 and 10 occasions (Table 7). In addition Troitskii and his associates carried out surveys of our neighbour galaxy M31, the Andromeda Nebula, to monitor any radioactivity originating with supercivilizations operating on a galactic scale (Kardashev's Type II).

The Soviet astronomers had introduced a special criterion to distinguish natural from artificial signals in their observations, carried out between October 1968 and February 1969: the more a signal resembled a pure sine wave, the more likely it was considered to be artificial. Troitskii reported that, within the accuracy of their instruments, no artificial signals had been observed. However, the Russian successor to the Ozma project applied also to signals which a civilization might unintentionally scatter into space.

At the Byurakan Conference, therefore, Troitskii also reported on the results of a search for pulsed broad-band radio emissions, such as would be accidentally released by the activities of extraterrestrial civilizations, for example in 'processing' huge amounts of energy.

Radio pulses are more easily detected, for the transmitter can propel them with an admittedly brief, but also high peak intensity. In March 1970, observations simultaneously began on three different wavelengths—50, 30, and 16 centimetres—at two locations: Gorki and the mountain Kara-Dag in the Crimea, 1500 kilometres away. At the very beginning of the operation the astronomers at the two telescopes simultaneously observed phenomena occurring on 50 centimetres. To clarify whether a purely statistical accident was causing the effect, it was decided to carry out the remaining programme at two additional observation posts: thus, as large an area as possible was covered by the four stations, guaranteeing the exclusion of any local terrestrial influences. The four telescopes, in Gorki, the Crimea, Murmansk and in the Ussuri district of Siberia, are separated by 1500 kilometres in latitude and by 8000 kilometres of longitude. The search for pulsed radio emissions was carried out in a rhythm of '2 days' observation, 2 days' rest' in the 2 months between 1 September and 12 November 1970.

The result was a clear confirmation of the first surveys: pulses of extraterrestrial origin were unquestionably recorded by all four stations, a far higher success rate than would be expected from a statistical accident. In general the pulses showed four basic patterns. The first group comprised short single impulses, some tens or hundreds of seconds long; in the second group the fluctuations increased like a stormy wind of fairly high strength; then, superimposed on these storms, a series of periodic impulses; and finally slow variations of the mean signal strength, without marked fluctuations.

The pulses were clearly observed physical processes which did not originate from Earth. But what did they prove? Were the pulses being unintentionally released by a distant civilization in the course of the explosive consumption of gigantic amounts of energy? It is all too easy for a laboriously built up scientific reputation to be destroyed if we prematurely turn to the sensation-seeking daily press with questionable announcements of success. A more critical self-questioning analysis finally led to the answer to the riddle: curiously enough, the supposed senders of the pulses had arranged to transmit them so that, in the main, they arrived at the Soviet observatories during the day. At night they occurred much less frequently. It had to be a process which manifested itself, by preference during daylight, within the atmosphere. The present interpretaion is that we are most probably dealing with a particular sort of solar activity. Troitskii finally noted drily: 'the experiment did not detect any sporadic emissions reaching us from the Galaxy.'[32]

But it seems that scientists, like anyone else, find it difficult to resist the temptation to report successful results. Two years later, in 1973, the Soviet news agency TASS announced once again that Troitskii and Kardashev had detected in Gorki radio signals 'of a type never received before'. Several times during the day a series of pulses was recorded for several minutes. This news again provoked mild excitement among Western scientists, for Samuel Kaplan, Director of the Research Institute in Gorki, announced that consideration was being given to an extraterrestrial origin for these strange signals. But this claim also fizzled out: soon afterwards it was announced that although these signals were definitely of artificial origin, they must originate in an area within the Solar System. Later it became clear that the signals almost certainly came from an Earth satellite, probably a secret military satellite from America.

In the 1970s further work was undertaken in the United States (see Table 8) to follow up the Ozma experiment—probably for reasons of competition. A fresh start was made by the South African radioastronomer Gerrit Verschuur back where it all began with Ozma in 1960: the National Radio Astronomy Observatory at Green Bank, West Virginia. Monitoring the 21-centimetre wavelength with two newly installed telescopes of diameter 46 and 100 metres, he observed 10 stars within 16 light-years of the Sun, among them of course the Ozma stars Tau Ceti and Epsilon Eridani. With the 100-metre telescope, Verschuur had the advantage of making observations with the most sensitive equipment available in the world at the time. While Verschuur was carrying out his main task of observing the neutral hydrogen clouds within the Galaxy, he repeatedly turned his telescope during October 1971 to Tau Ceti, Epsilon Eridani and 61 Cygni, and in August 1972 he turned the 46-metre telescope to 10 other stars. Verschuur estimated that the sensitivity of his equipment so far exceeded that of Drake's Ozma receiver of 11 years earlier that 5 minutes of his observing time corresponded to about 4 days' observation by the Ozma project.

Verschuur's observations of radio waves from Tau Ceti and 61 Cygni surpassed Troitskii's 1969 results in sensitivity by a factor of 1000. Despite this, no artificial signals were traced from the stars monitored by Verschuur.

What is probably the most thorough Ozma-style investigation of Sun-like stars has been carried out since 1972 by the American astronomers Ben Zuckerman and Patrick Palmer, who recorded on magnetic tape the radio signals of about 650 stars within a radius of 75 light-years of the Sun. Each star was observed for half an hour, with special attention given to signals varying in a rhythm of minutes or days and/or covering a very narrow bandwidth. Just 24 stars showed suspicious fluctuations in their radio noise (personal communication, 1977). Zuckerman and Palmer intended to examine these stars further.

Ozma already lies far in the past. Since then, advances in radioastronomy and above all the desire for a systematic search for artificial extraterrestrial signals have led to new and almost utopian projects. One of these will be described in the following section.

Table 8 Projects to search for extraterrestrial civilizations

Date	Research group	Observatory	Search frequency (gigahertz)	Resolution (hertz)	Limiting sensitivity (watts per m²)	Total hours	Size of telescope (metres)	Targets
1960	Drake: 'Ozma'	NRAO	1.420–1.4204	100	4×10^{-22}	400	26	Epsilon Eridani; Tau Ceti
1968-69	Troitskii, Rakhlin, Starodubtsev & Gershtein	Zimenkie, USSR	0.926–0.928	15	2×10^{-22}	11	55	12 nearest Sun-like stars
1970 and present	Troitskii, Bondar & Starodubtsev	Eurasian Network, USSR	1.863, 0.927 and 0.600			700	Dipole (isotropic)	Pulsed signals from whole sky
1971-72	Verschuur: 'Ozpa'	NRAO	1.4198–1.421 and 1.410–1.430	490 and 6900	5×10^{-24} and 2×10^{-23}	19	91 and 43	9 nearby stars
1972-76	Palmer & Zuckerman: 'Ozma II'	NRAO	1.415–1.425* and 1.4201–1.4207	64000 and 4000	1×10^{-23} and 3×10^{-24}	~500	91	~650 Sun-like stars within 75 light-years
1972	Kardashev	Eurasian Network; Institute for Cosmic Radiation	1.337–1.863				Dipole	Pulsed signals from whole sky
1973–	Dixon & Cole	Ohio	1.4204	20000 to 100000	1.5×10^{-21}	Continuous	Dipole	Entire sky
1974–	Bridle & Feldman	ARO	22.2	30000			46	500 nearby stars
1974–	Wischnia	Copernicus satellite	Ultraviolet					Epsilon Eridani, Tau Ceti, Epsilon Indi

1975–76	Drake & Sagan	Arecibo (Puerto Rico)	1.42, 1.653 and 2.38	1000	3×10^{-25}	~100	305	Type II civilizations in 4 nearby galaxies
1976–	U. of California, Berkeley: 'Serendip'	Hat Creek, California	1.41, 1.853 and 0.917	2500	5×10^{-22}		26	Entire sky
1976	Clark, Black, Cuzzi & Tarter	NRAO	8.522–8.523*	5	2×10^{-24}	7	43	VLBI of 4 stars
1977	Black, Clark, Cuzzi & Tarter	NRAO	1.665–1.667*	5	4×10^{-25}	100	91	200 stars
1977	Drake & Stull	Arecibo (Puerto Rico)	1.664–1.668*	2.5 (best)	8×10^{-26} (greatest)	10	305	6 Stars
1978	Edelson (JPL, Pasadena)	Goldstone, California	1.4–25	300	3×10^{-23}		26	Entire sky
1978	Clark, Black, Tarter, Cuzzi & Conners (ARC)	N and S Hemispheres, Arecibo	1.4–1.73	10 and 0.01				Selected stars within 1000 light-years

* These observations were corrected for Doppler-shifted frequencies caused by motions of the stars.

NRAO is the National Radio Astronomy Observatory in Green Bank, West Virginia; Ohio denotes the Ohio State University Radio Observatory (Columbus, Ohio); the Eurasian Network is formed by collaboration between telescopes in Gorki, the Crimea, Murmansk and Ussuri directed principally by V. S. Troitskii of Gorki University together with N. S. Kardashev of the Institute for Cosmic Research of the Academy of Sciences of the USSR; ARO is the Algonquin Radio Observatory in Algonquin Park, Canada; Arecibo is the 305-metre radio telescope of Arecibo Observatory in Puerto Rico; Goldstone (California) possesses a number of radio telescopes, of which the largest is 26 metres in diameter; JPL is the Jet Propulsion Laboratory in Pasadena, California; ARC is NASA's Ames Research Center at Mountain View, California. VLBI is Very Long Baseline Interferometry.

The two studies planned at JPL and ARC deal with the problem of selecting a frequency in a way which distinguishes them from earlier projects of this nature: they make use of what is known as a multi-channel frequency analyser. With this instrument measurements can be made at the same time on a large number of frequencies ('channels'), so that an extended band of frequencies can be monitored simultaneously—and with fine frequency resolution on each channel. The new analyser is intended to process over a million channels. (Table partly compiled from ref. 33.)

Projects of today: Cyclops

Interim report from the year 2195:

It had been conceived in the first bright dawn of the space age: the largest, most expensive, and potentially most promising scientific instrument ever devised. Though it could serve many purposes, one was paramount—the search for intelligent life elsewhere in the Universe.

One of the oldest dreams of mankind, this remained no more than a dream until the rise of radio astronomy, in the second half of the twentieth century. Then, within the short span of two decades, the combined skills of the engineers and the scientists gave humanity power to span the interstellar gulfs—*if* it was willing to pay the price.

The first puny radio telescopes, a few tens of metres in diameter, had listened hopefully for signals from the stars. No one had really expected success from these pioneering efforts nor was it achieved. Making certain plausible assumptions about the distribution of intelligence in the Galaxy, it was easy to calculate that the detection of a radio-emitting civilization would require telescopes not dekametres, but kilometres, in aperture.

There was only one practical method of achieving this result—at least, with structures confined to the surface of the Earth. To build a single giant bowl was out of the question, but the same result could be obtained from an array of hundreds of smaller ones. *Cyclops* was visualized as an antenna 'farm' of hundred-metre dishes, uniformly spaced over a circle perhaps five kilometres across. The faint signals from each element in this army of antennae would be added together, and then cunningly processed by computers programmed to look for the unique signatures of intelligence against the background of cosmic noise.

The whole system would cost as much as the original Apollo [Moon landing] Project; but, unlike Apollo, it could proceed in instalments, over a period of years or even decades. As soon as a relatively few antennae had been built, Cyclops could start operating; from the very beginning, it would be a tool of immense value to the radio astronomers. Over the years, more and more antennae could be installed, until eventually the whole array was filled in; and all the while, Cyclops would steadily increase in power and capability, able to probe deeper and deeper into the Universe . . .[34]

We have not yet come as far as the scheme described by the British writer Arthur C. Clarke in his science fiction story *Imperial Earth*, looking back from the twenty-second century. At present, Project Cyclops is no more than a study made in 1971 by NASA scientists to evaluate the technical limits of a system for the discovery of extraterrestrial intelligent life. A quotation from Frank Drake is prefaced to the investigation:

At this very minute, with almost absolute certainty, radio waves sent forth by other intelligent civilizations are falling on the Earth. A telescope can be built that, pointed in the right place and turned to the right frequency, could discover these waves. Someday, from somewhere out among the stars, will come the answers to many of the oldest, most important and most exciting questions mankind has asked.[35]

To turn such bold visions easily into reality requires that many civilizations within the Milky Way, for instance, are actually making systematic attempts to establish interstellar contact with continuous sending and search programmes along the lines of the strategies described above. For if we ourselves were the addressees of the Arecibo message of 1974, this 3-minute transmission could easily flash past the Earth unnoticed if it reached us at this moment.

The chief concern of the communications engineers and astronomers at NASA's Ames Research Center was the solution of technical problems, such as how to build receivers steerable by remote control, with variable frequency and a central signal processing system. The study concludes that it is already technically possible to construct a super-receiver of this kind with a total receiving surface greatly exceeding anything available today (Figure 43). The idea of the Cyclops system—instead of one large telescope dish to construct hundreds of smaller, centrally steerable dishes—has the advantage that even during its construction it can be used with the greatest flexibility. The Cyclops arrangement can be enhanced with extra dishes almost at will, but if 'results' already appear before the planned completion, the construction can be stopped earlier.

Cyclops is versatile, too, in its possible applications. The receiving area of the system can be divided at will into smaller, separately operated systems to carry out special scientific programmes.[36] The cost of a circular Cyclops installation 10 kilometres in diameter is at present estimated at 1200 million US dollars. This system would then correspond to a single gigantic aerial dish 3 kilometres across.

As planned by the engineers, Cyclops would operate automatically most of the time, following its assigned tasks, and only call on human operators in the event of an anomaly. Only by such methods could the necessary study of the heavens over long intervals be carried through for decades at a time, and the scientific frustration at negative results be kept bearable. Bernard M. Oliver, one of the directors of the Cyclops group, envisages the final installation as a plantation of antennae covering 20 square kilometres, feeding all their data into a single data processor and all under automatic computer control.[36] Each telescope would be as large as the radio telescope in Bonn-Effelsberg, currently the largest steerable radio telescope in the world (Figures 44 and 45).

The Cyclops team plan the construction of the system in four phases. Even before the first aerial dishes and the computer system are put in operation, the list of target objects should be completed. There is still no complete catalogue of the most likely targets within 1000 light-years of the Sun. In other words, we do not know, apart from the few hundred nearest stars studied by optical astronomy, where within a distance of 1000 light-years the stars which offer the best chances for the origin of life are to be found. The aim of the first phase, then, is to sort out the stars most like the Sun by ordinary optical means, so that no valuable Cyclops time will later be squandered on listening to stars which, according to NASA, do not offer the slightest probability of civilizations with radio technology.

Figure 43 The complete Cyclops installation would probably cover 20 square kilometres. The radio telescopes, each 100 metres across, would be steered by a control system at the centre of the circle. (Courtesy: NASA.)

In seeking places where life has arisen, this 'short-listing' of stars may be well advised; for it is probably true that intelligent life forms principally arise on planets. On the other hand, it is most questionable whether intelligent life, especially if it has developed a high technology, will continue to *occupy* only planets. If, for instance, expansion into space colonies (rather than the colonization of planets) is characteristic of technological civilizations, the colonists might select stars according to how well they fulfilled their requirements of energy and raw materials rather than whether new life could originate in their vicinity. In this case, the hotter and shorter-lived stars belonging to spectral classes B, A, and F might also be considered.

Given the erection of about 100 aerials a year, the second phase, the actual construction, will require from 10 to 25 years. The search, however, can very quickly be put into operation. If each star is observed for half an hour, Cyclops would very soon get through about 15 000 stars a year. Up to the time when Cyclops is completed, the stars up to a distance of 500 to 700 light-years can perhaps be run through twice. The complete installation will finally make it possible to advance to 1000 light years (phase three). At the same time the plan includes sending artificial signals in the style of the Arecibo message repeatedly at regular intervals, again examining the nearer stars after some decades of operation for answering signals.

If by the end of the third phase no artificial signals have been detected, the programme, hitherto principally devoted to searching, may need to be considerably modified. By then the list of potentially life-supporting stars will be complete and can be subjected to further automatic checks; longer messages can thus be sent more often (phase four). The seekers for contact, of course, will always have the worry that *all* the civilizations concerned may have settled exclusively for searching, sitting mute at their radio and other telescopes and patiently examining the heavens. But the Cyclops team hopes that this phase will never be reached in their search, since other races will already have reached phase four.[35] In New Mexico in the United States a forerunner of Cyclops, the so-called Very Large Array, has now been in operation for a few years; it is still being extended.

Even science fiction writers estimate the chances of success with super-telescopes like Cyclops rather pessimistically. In 1972 James E. Gunn described in his novel *The Listeners*[37] the situation where radioastronomy is only carried out via telescopes in space, and the Arecibo telescope is free to search exclusively for artificial signals. In Gunn's story a team searches the skies for 50 years until finally—in the year 2028—'success' results. And what does the decoded message

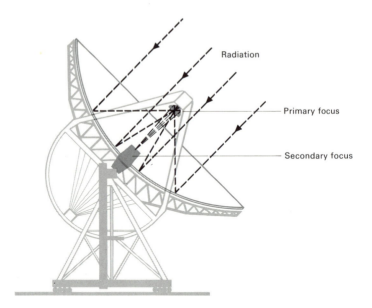

Figure 44 Diagrammatic view of the steerable radio telescope in Effelsberg, near Bonn, W. Germany. The incoming radio waves are collected by the parabolic surface to the primary focus, where they are reflected by a mirror to the central point of the telescope's reflector, the secondary focus. Thirty metres above the base of the dish is a cabin containing the receiving system.

show? A Drake-style pictogram showing a bird-like creature from a solar system in the constellation Capella, whose star is about to flare up as a supernova!

Earthly optimism alone will not make extraterrestrial civilizations more enthusiastic about sending messages, even though (starting from the assumption that the origin of life is a normal process) the hypothesis that most civilizations at some time in their development will recognize the likelihood of life on other stars is basic to the Cyclops project. As soon as they are technically in a position to initiate search and sending projects, therefore, they will do so, and thus draw attention to themselves and eventually be discovered (see Part III: 'The Galactic Club', page 235).

While American astronomers are still dreaming about Cyclops, a giant radio telescope was constructed for the Soviet Academy of Sciences in 1976 in the northern Caucasus. With an effective diameter of 600 metres, it is twice as large as the radio telescope at Arecibo, hitherto the largest, and carries the name Ratan 600.[38] The structure has an advantage over Arecibo in its tracking facility, by which target objects can be kept in view longer than with fixed telescopes. This is achieved by dividing the parabolic dish into 895 separate mirror surfaces, each

Figure 45 The 100-metre diameter radio telescope in Bonn-Effelsberg, is the largest fully steerable telescope in the world. (Courtesy: Max-Planck-Institut für Radioastronomie, Heidelberg.)

2 by 7.5 metres and rotatable about two axes. The simultaneous rotation of the mirrors is governed by a computer. They can also be connected in four separate sections, each of which can be operated as a single telescope with an effective diameter of 300 metres.

The astronomy of visible light, too, is in a state of technical revolution. Near Ratan 600, there appeared in 1976 an optical telescope 6 metres in diameter.[39] Until then, the world's largest telescope had been the Hale 5-metre telescope on Mount Palomar in the United States. 'I think that [the Soviet 6-metre telescope] is the largest of all optical telescopes that will ever be built on the Earth, for it goes right up to the limits set by the Earth's atmosphere on the observation of the Universe. The next step can only be their installation on platforms orbiting the Earth,' said Bernard Lovell, Vice President of the International Astronomical Union. In fact NASA has long been planning to launch an optical reflector into Earth orbit with the Space Shuttle.

But even among Earth-bound optical telescopes a revolution is beginning.

While the telescopes on Mount Palomar and in the northern Caucasus certainly represent almost the technological limits of single optical mirrors, astronomers at the Kitt Peak National Observatory near Tucson in Arizona, USA have developed plans (see Figure 46) for an optical telescope with an effective size of 25 metres.[40] Work with multi-mirror systems has already shown that it is possible to align them so precisely by reflected laser beams that all the mirrors direct the light to one focus. Also, it is possible by using a special technique of electronic image enhancement—known as speckle interferometry—to remove the damage done by atmospheric turbulence. By these means, the resolution accuracy of optical instruments can be considerably increased.

A 25-metre telescope would be assembled from a large number of different mirrors. It could observe objects in the sky which have hitherto been beyond our reach optically, simply because they are too faint for existing telescopes; planets around nearby stars, dark dwarf stars, the motions of stars in nearby galaxies, and variations in weather on the other planets in the Solar System.

Radio contact between galaxies

Despite their great sensitivity to incoming signals, telescopes of the Cyclops type are restricted in range to our own Galaxy—except where an actual supercivilization is involved. Radio signals from these would be readily observable in neighbouring galaxies too.

In normal conditions, however, radio contact between civilizations in different galaxies introduces quite special considerations. If we assume the existence in every galaxy of at least one civilization sending and receiving information, it is certainly justifiable to examine neighbouring galaxies for artificial signals with directional transmitters and receivers. The angle covered by the aerial in such a case need only be large enough to include the whole of the galaxy under examination. The choice of a target for the Arecibo message of 1974 was made on this principle: the globular star cluster M13, 24 000 light-years away, is exactly covered by the beam of the Arecibo instrument (angular size 0.033 degrees at 2.3 gigahertz). The more distant the target galaxy, the more the signal can be concentrated (beamed), the greater is the directional factor which can be introduced. This circumstance even has a more favourable effect on the required transmission power, because the increase in strength made necessary by the greater distance rises more slowly. As Vsevolod Troitskii pointed out at the first Byurakan Conference on extraterrestrial civilizations in 1964, this leads, for intergalactic distances, to the following paradox: the greater the distance between communicating galaxies, the lower is the transmission strength necessary to establish communication.

This paradoxical effect is illustrated in Table 9 for several nearby galaxies such as the Magellanic Cloud and the Andromeda Nebula, and also a typical galaxy of

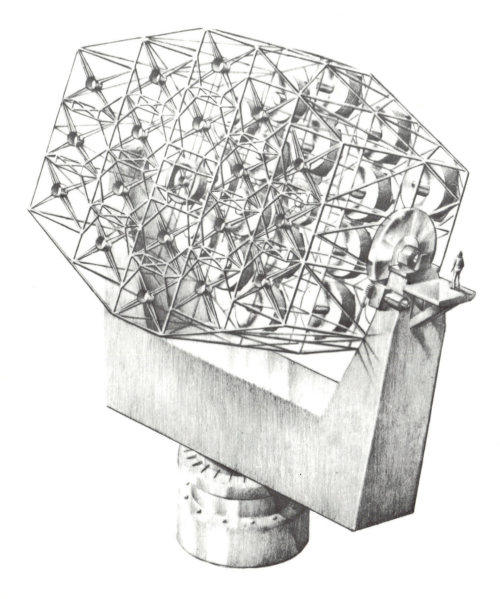

Figure 46 Astronomers at the Kitt Peak National Observatory near Tucson, Arizona, USA, hope to attain an effective mirror diameter of 25 metres with this super-telescope composed of several hundred individual reflectors, shown here in a design sketch. At present the Soviet 6-metre telescope in the northern Caucasus is the largest optical reflecting telescope in the world. (Courtesy: Kitt Peak National Observatory.)

Table 9 Transmission strength and aerial size for communication with other galaxies (optimum conditions)

Galaxy	Large Magellanic Cloud	Andromeda	Galaxy of same size as our own			
Distance in light-years	2×10^5	2×10^6	10^7	10^8	10^9	10^{10}
Angular size of target galaxy	$9°$	$3.5°$	$34'$	$206''$	$21''$	$2''$
Dish area needed for wavelength of 3.5 centimetres in square metres	0.04	0.3	10	10^3	10^5	10^7
Transmission strength in watts	6×10^{23}	10^{24}	2×10^{22}	2×10^{20}	2×10^{18}	2×10^{16}

For comparison: diameter of our Galaxy = 10^5 (100 000) light-years. In calculating the transmission strength, an optimum directional strategy is assumed for both transmitter and receiver: they are assumed to have equally good directional aerials and to use the same frequency band only 1 hertz in width. The location of the hypothetical transmitter is always somewhere within a previously selected galaxy. Beyond the distance of the Andromeda Galaxy, the transmission strength required falls with increasing distance (given these optimum conditions). ($°$ = degrees; $'$ = minutes of arc; $''$ = seconds of arc; 60 seconds of arc = 1 minute of arc = 60 degrees.) (After ref. 21.)

the size of the Milky Way at various increasing distances. In the extreme case of infinite distance, it appears that a radio telescope would require practically no transmission power at all, and could still transmit messages halfway across the Universe. Needless to say, this is a nonsensical result. The approximate relationship between increasing distance and diminishing signal strength over intergalactic dimensions is strictly valid only for the typical distances of certain galaxies, and should not be extrapolated too far. The upper limit is reached when the beam from the aerial can no longer be made any narrower (the upper limit of the directional factor) for either transmitter or receiver. If it is nevertheless desired to home in on more distant galaxies, the directional beam will cover a larger and larger area of the sky compared with that occupied by the galaxy itself.

Most of the search programmes hitherto completed have principally analysed nearby stars like the Sun. But quite recently some researches have been undertaken to survey a number of galaxies for artificial signals.[41] The radioastronomers Alan Bridle and Paul Feldman, with James Condon of the Virginia Polytechnic Institute, have suggested examining about 150 nearby galaxies around the wavelength of the hydrogen line with the 300-metre Arecibo telescope. The three astronomers envisage that the experiment would use Arecibo's new 1000-channel analyser, which breaks down the frequency band to be investigated into 1000 narrow frequency channels: The astronomers thus attain an accuracy in frequency resolution of 150 hertz per channel, which permits the separation of artificial signals with the expected narrow frequency bands from the broad-band noise of interstellar clouds. According to Feldman: 'The unique feature of the experiment is that by pointing at galaxies, approximately 10^{11} stars are in the beam.'[41]

At Easter 1975 Frank Drake and Carl Sagan undertook independently a more modest version of this programme, in which they studied four nearby galaxies with the Arecibo telescope. These galaxies are among the approximately 20 members of the 'local group' of galaxies, to which the Milky Way also belongs.

What of the transmission energy which must be provided to bridge the enormous distances between galaxies? How far can the range for interstellar and intergalactic communication be extended? This depends above all on the energy available. But how much energy can a civilization afford to radiate away? On Earth the situation is that during the past century energy consumption has increased every year by an average of 3 per cent. If we naively extrapolate this curve, in 300 years' time the energy expanded reaches the value of 10^{17} watts, which matches the solar radiation reaching the Earth. In practice much less energy will certainly be consumed 300 years from now.

This is because, according to all we know about the Earth's climate, we shall have to 'freeze' the energy consumption at a much lower level if mankind is not to risk the destruction of its biosphere, or drastic changes of climate at the least. A turnover of energy of even a few per cent of that radiated by the Sun (1.74×10^{17} watts) would probably already be catastrophic.

An example is supplied by carbon dioxide: the burning of fossil fuels, coal, oil and natural gas, liberates this gas, which settles above the atmosphere like a filter. The carbon dioxide layer admits (short-wavelength) light to the Earth, but does not allow the returning (long-wavelength) radiation back into space. This greenhouse effect (see Part I: 'The life-supporting zones of the stars', page 32) causes the average temperature to rise. An increase of temperature by an average of as little as 1 °C would already have serious consequences: the formation of deserts, raising of the sea-level by several metres, catastrophic floods in all coastal regions, loss of living space.

On the other hand, a drop of no more than 5 °C in the mean annual temperature would set off a new ice age. Conversely, the mean temperature can only rise by perhaps 0.5 °C and still be climatically bearable. With a 7 per cent annual energy increase, we should have reached this limit in only 82 years.[42] The energy consumed every year would then represent 1.2 per cent of the radiation received from the Sun, and 317 times the present consumption of 6.6×10^{12} watts! So all our problems cannot simply be solved by 'more energy'.

This utopia, which could be attained by mankind within a few decades, is only outlined here to illustrate that a maximum energy consumption on the order of 10^{15} watts is the most we can allow ourselves if we are not to destroy the sensitive equilibrium of our terrestrial climate.

It must be emphasized that this upper limit for *planetary* civilizations

Table 10 Transmission ranges of different types of civilization in unfavourable conditions (isotropic transmission and reception)

| Civilization | Range of signals | | Frequency bandwidth of artificial signals |
	COMMUNICATION (Reception of information)	DISCOVERY (Evidence of existence)	
Type I		A few light-years	More than 1 hertz
Type II	A few light-years A few thousand light-years	~ 1000 light-years Galaxy	1 megahertz 100 hertz (monochromatic)
Type III	A few nearby galaxies Clusters of galaxies (a few thousand galaxies)	Known Universe	10 megahertz 100 hertz (monochromatic)

(Kardashev Type I) is probably universally valid. It is not a matter of the limited availability of raw materials, but follows directly from the physical necessity of maintaining thermodynamic equilibrium in the Earth's atmosphere. Civilizations restricted to planets must all be more or less subject to this limitation.

What transmission energies can supercivilizations afford? Civilizations can only permit themselves consumption of energy above the estimated 10^{15} watts if they extend their industrial zones to the space between the planets of their system, and beyond it. What time intervals must be taken into account then? Let us assume a growth in energy consumption of 1 per cent. A planetary society (Type I) would then reach the energy turnover of a Type II civilization (3×10^{26} watts) in a few thousand years. If the development continues at this pace, after a further 2500 years energy would be consumed in similar quantities to those emitted by an average galaxy in the form of radiation (10^{37} watts). By Kardashev's classification this society would then join the club of galaxy manipulators, Type III civilizations. By choosing suitable frequency bands it could make itself noticed throughout the whole of the known Universe (Table 10).

These conclusions are naturally valid under the assumption that civilizations of all Kardashev types, when they consume more energy, also radiate correspondingly more energy into space. This is what is happening on Earth at present; but it is not at all clear whether things will continue in the same way. For another trend is already setting in: alongside the technology of television stations and television aerials, the age of cable television has long since dawned. The principle of energy saving can very quickly lead highly developed civilizations, in other technological fields too, to prevent the unintentional squandering of energy by radiation into space, or to reduce it to a minimum. Consequently, no more radio transmissions would reach the stars without intention: radio silence will reign for all eavesdroppers in the Galaxy. On the Earth, this stage may be reached within a few decades, and the period of unintentional radio transmission thereby limited to about 100 years—the twinkling of an eye on the astronomical timescale of galactic communication.

Contact with extraterrestrial civilizations by light and laser

There are many good reasons for using radio waves as the means of exchange of information between the civilizations of our Galaxy and other galaxies. Let us briefly summarize here the arguments that have been advanced in earlier sections. The level of cosmic noise falls to a minimum in the microwave window; radio waves from the directional transmitters available today will already reach the greater part of the Galaxy—radio waves penetrate the dust in the equatorial plane of the Galaxy almost unhindered—and the Sun also radiates relatively weakly in the centimetre range. Artificial radio signals stand out well against the thermal radio noise from the Sun. But there are disadvantages too. Contrary to the original promise of the 21-centimetre line, we must choose definite wavelengths for observation more or less arbitrarily.

The second astronomical window, visible light, comes off worse in many ways. Light is quickly absorbed by interstellar dust. In the direction of the galactic plane the intensity of a light source is diminished by a factor of 100 per kiloparsec (3260 light-years). In these directions, the Milky Way is only transparent again at wavelengths larger than 50 to 60 micrometres, that is, beyond the infrared and approaching the radio region, if we wish to observe the whole Galaxy.

If we limit ourselves to smaller distances, however, to stars up to a few thousand light-years away, for instance, the absorption of visible light no longer plays such an important role. But any optical artificial signal still risks being overwhelmed by the background of solar radiation, for the Sun radiates most strongly in exactly this range.

The technical developments of the last 20 years (especially in *masers* and *lasers*) and the possibility of setting up telescopes on satellites outside the Earth's atmosphere have, however, rendered optical signals, with the neighbouring frequencies in the infrared and ultraviolet included, rather more attractive for the purposes of interstellar communication.

Lasers are amplifiers of radiation, which can send out extremely 'clean' light—that is, signals of very small bandwidth—in a narrow beam. The word laser is an acronym derived from the initials of '*L*ight *a*mplification by *s*timulated *e*mission of *r*adiation'. When applied to microwaves, the same principle has led to very sensitive transmitters, the *masers*. The word maser stands for '*m*icrowave *a*mplification by *s*timulated *e*mission of *r*adiation'. The first lasers were made only from ruby crystals. Nowadays, lasers are routinely constructed from semiconductors, glass, liquids and gas mixtures, which emit their light in a great variety of wavelengths from the infrared to the ultraviolet (see Part III: 'To the stars with laser fusion', page 248).

Above all, it is the sharp directional beaming of laser radiation that has made optical light sources interesting, at least for interplanetary purposes. At first sight it might seem sufficient to direct a powerful searchlight at the sky. But the inescapable fanning out of the light beam in an ordinary searchlight makes this attempt useless. Modern projectors have an angular aperture of at least some 30 minutes of arc. If this beam of light is directed to the Moon, it already covers an area measuring about 3080 kilometres on the Moon's surface. If we take a lamp 100 kilowatts in power, it would only be visible from the Moon as a faint star of the fifteenth magnitude—comparable to the brightness of the little satellites of Mars as seen from the Earth.

It is different with lasers. The light rays are so nearly parallel when they leave the 'optical resonator'—with an aperture of only 1 to 10 seconds of arc, two-hundredths of that of an ordinary light ray—that the light beam would have a diameter to only a few kilometres on the Moon. Astronomers have already used this property to measure the Earth–Moon distance with laser rays more accurately than ever before. During an Apollo mission, astronauts set up mirrors on the Moon specially constructed to reflect laser beams as efficiently as possible. On precise measurement of the time taken by a laser pulse for one round trip from the Earth to the Moon and back, the distance could be accurately determined to within metres.

The other important property of the laser for interstellar communication is the small bandwidth of the beam. The beams of some lasers built today have a bandwidth on the order of 1 kilohertz. This is a hundred-thousand-millionth (10^{-11}) part of the frequency of the optical signal itself. An incandescent light radiates waves from the infrared over the whole visible spectrum, corresponding to a bandwidth of some terahertz (10^{15} hertz).

There are also lasers which radiate in the ultraviolet. NASA is now looking for artificial ultraviolet laser signals. The research station is the NASA satellite OAO 3 (Orbiting Astronomical Observatory), also known as Copernicus, which has been orbiting the Earth since 1972 at a distance of 750 kilometres. With a 77-centimetre spectrometer mounted on the satellite, the three Ozma veterans—the Sun-like stars Epsilon Eridani, Tau Ceti and Epsilon Indi—have been observed since 1974 to see whether unusual ultraviolet laser beams indicate the existence of highly developed life-forms.[43]

But hitherto the search has been fruitless. In order to make a realistic estimate of the possibilities of lasers, the Cyclops project group at NASA has undertaken a detailed comparison between six different laser systems—two in the optical, two in the infrared, and two microwave systems. We shall consider this comparison more deeply because lasers do not turn out to be as well suited for interstellar purposes as one might expect. The first to consider how lasers might be applied to interstellar communication were the American physicists R. N. Schwartz and C. H. Townes.[44] Townes himself had made important contributions to the development of masers—he received the Nobel Prize for it—and had also discovered the first molecule in space with four atoms, ammonia, in 1968 (see Part I: 'Prebiotics in interstellar space: from hydrogen to hexose', page 11). The Cyclops team based itself on their ideas; here we give the principal results of their analysis.

The question is: can a laser compete in range with a microwave transmitter (a radio telescope)? (Microwaves in the centimetre range are also received by radio telescopes with special equipment.) The question was investigated for a micro-wave system (system A) and for laser systems which can be made today, but also for a laser system which may only be technically possible in an improved future (system B). Two lasers, optical and infrared, were now compared with a microwave transmitter. For both transmitter and receiver of system B, the angular aperture of the beam was taken to be only 1 second of arc. However, microwave system A has an angular aperture of 64 seconds of arc, corresponding to the present state of the art. Nevertheless, microwave system A beats both the laser systems in range by a factor of 200. And microwave system B, with the same beam aperture as both laser systems, covers a territory 10 000 times greater. With a total transmitting power of 10 megawatts, microwave system B can even send signals over *intergalactic* distances. Even infrared laser systems—carbon dioxide gas lasers radiating at 10.6 micrometres—and brought in by Cyclops for additional comparison, do no better than neodymium-glass lasers.

Why do lasers do so poorly in comparison with microwaves? The only advantage of lasers, that they produce narrow light beams with small transmitters, becomes a disadvantage for the receiver, since he requires large surfaces for his receiving aerials.

• Antenna surfaces are cheaper and more durable for microwaves, since tolerances can be greater at these longer wavelengths, and polished surfaces are not required.
• Microwaves have larger ranges for the same transmission power, even with a greater angular aperture, which again makes the aerial easier to construct.
• Microwave aerials are all-weather systems.
• Doppler shifts, frequency variations and system noise are smaller by orders of magnitude than for laser signals.

But at short distances—say, within a distance of 50 light-years—laser communication still offers an alternative. Nevertheless: 'If lasers had already been known for centuries and microwaves had only recently been discovered, microwaves would be hailed as the long-sought answer for interstellar communication,' comments Bernard M. Oliver.[45] Only if fusion research succeeds in increasing laser intensities a millionfold will this situation change. We shall become acquainted later with a totally different possibility, that of discovering evidence of other civilizations in the infrared light of the stars (see 'The astroengineers of supercivilizations', page 200).

X rays and extraterrestrials

This section on X rays, along with the following ones dealing with gravitational waves, neutrinos and tachyons, is much more speculative in character. Their subject is not quite science fiction, but only in the case of X rays are we at present dealing with an experimental verified physical reality. Although tachyons and gravitational waves (it will annoy many physicists that both are named in the same breath) are accepted in theoretical physics as *possible* realities—gravitational waves far more than tachyons—no positive evidence of the existence of either phenomenon has hitherto been yielded by experiment.

It is different with X rays. Since Wilhelm Conrad Röntgen's accidental discovery of this phenomenon in Würzburg in 1895, they have been applied with the greatest success in medicine, biology and physics. A special branch of astronomy, too, X-ray astronomy, has developed rapidly since satellites discovered X rays from space, which are intercepted by the atmosphere. The first measurements of extraterrestrial X rays—which are only observable from above about 40 kilometres—were made from rockets in 1948, and later also with balloons. But it was satellites which started the 'boom'.[46,47]

On Kenya's Independence Day, in December 1970, NASA launched the first American–Italian Uhuru satellite off the coast of Malindi in Kenya, into an orbit above the Earth's equator. ('Uhuru' is the Swahili word for 'freedom'.) Within a short time Uhuru had detected 160 X-ray sources, of which probably about 100 belong to the Milky Way. But only about 10 per cent of these sources could be identified with previously known types of stars—such as remnants of supernova explosions. In other sources some then unknown mechanism had to be producing the energy radiated as X rays, which in some cases exceeds the total radiation of the Sun by a factor of almost 100 000 (10^5).

Especial interest was aroused by sources emitting X-ray flashes at exact periodic intervals of a few seconds (Figure 47). The periodicity of X rays was first discovered in 1971 in the source Centaurus X-3. Today we are confident that this and similar systems are double stars in which something like a rotating neutron star with a strong magnetic field is in orbit about a giant star. The giant is just at the stage of its development in which it is expanding, so that gas is streaming towards the small, but dense neutron star. The observed X rays are produced as the gas falls on to this neutron star along its magnetic field (Figure 48). The rapid rotation of the neutron star then sends us a short burst of X rays with each rotation, like a lighthouse.

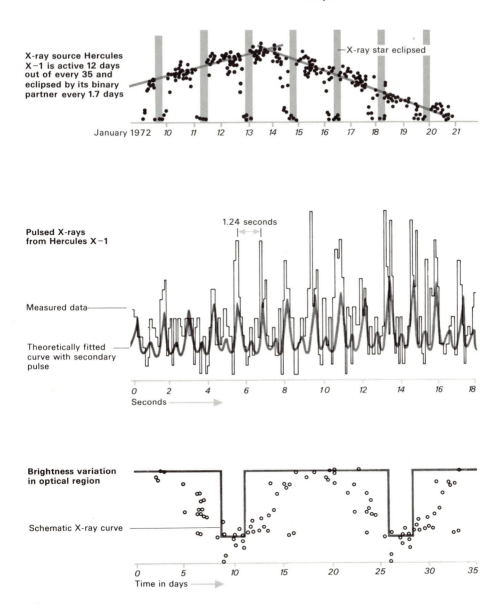

X-ray source Hercules X−1 is active 12 days out of every 35 and eclipsed by its binary partner every 1.7 days

—X-ray star eclipsed

January 1972 10 11 12 13 14 15 16 17 18 19 20 21

Pulsed X-rays from Hercules X−1

1.24 seconds

Measured data

Theoretically fitted curve with secondary pulse

0 2 4 6 8 10 12 14 16 18
Seconds

Brightness variation in optical region

Schematic X-ray curve

0 5 10 15 20 25 30 35
Time in days

Figure 47 Data on the X-ray source Herculis X-1 are derived from measurements in different regions. The top curve shows the X-ray emission of the source during a 12-day phase of activity in 1972. The continuous line shows a sudden onset of the radiation on 9 January. The break in transmission every 1.7 days, when the neutron star is eclipsed by its partner for 5.7 hours, can be clearly seen. The X rays from Herculis X-1 show pulses matching the 1.24-second rotation period of the neutron star (centre). The bottom curve shows the optical brightness variation of the optical star HZ Herculis, identical with Herculis X-1.

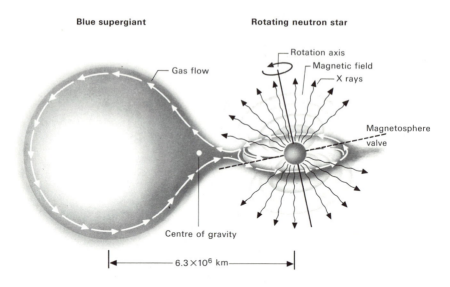

Figure 48 The gas flowing between the elements of the binary star Herculis X-1, HZ Herculis emits X rays as it falls on the neutron star.

Nature employs massive binary systems to produce X rays which reach the Earth across great distances. But human technical facilities are not yet far enough developed for us to emulate Nature's binary star process, to juggle with stars so as to convert them into transmitting stations for X rays (or other kinds of radiation). The nearest approach to this astrophysical situation which can be achieved in terrestrial circumstances lies in nuclear explosions.

These would do no damage to the Earth if they were detonated outside the atmosphere—a minimum height of 80 kilometres would suffice. Today it is technically possible to put a series of hydrogen bombs into Earth orbit, and to detonate them according to a definite code, like a Morse message, at regular intervals (not, perhaps, the worst possible use for our stock of atomic bombs). Such a bombing message was in fact already sent to the stars at the beginning of the 1960s—though not exactly intentionally. In this era of nuclear testing the United States detonated at least five atomic bombs in the upper atmosphere, two at least in the megaton range. If we assume an equal number of nuclear tests for the Soviet Union, about 10 powerful X-ray pulses were sent into space from our planet at that time. But up to what distance could they be observed?

In the explosion of an atomic bomb (with a duration of 1 microsecond), a considerable part—about 70 per cent—of the energy is liberated in the form of X rays.

On the face of it, X rays from nuclear explosions have useful properties for communication purposes. This short-wave radiation is hardly affected by the interstellar medium (unlike long-wave radio radiation). Besides this, the X-ray flash covers a wide frequency band, and stands out clearly against the weak X radiation of the Sun.[48]

In 1 microsecond, our quiet Sun radiates only a ten-thousandth (10^{-4}) of the energy of the bomb as X rays, with a power between 1 and 10 kilovolts. Even a solar flare coinciding with the detonation of the bomb will produce at the most a hundredth (10^{-2}) of the nuclear X-ray intensity. But the stumbling block for this useful application of nuclear weapons is the most important consideration: the range. The communication range is rather disappointing. The X rays from a 1.4 megaton bomb, which produces 3×10^{22} ergs as X radiation, could only be certainly detected with our present terrestrial receivers up to 400 astronomical units away. This represents 10 times the distance from the Sun to the outermost plant in the Solar System, Pluto, about 2 light-days. And what is the *maximum* distance attainable? If we put together *all* the atomic bombs of the United States and the Soviet Union, each about 10 000 megatons, the distance is, of course, increased. If in addition we could beam the X rays in one direction with an angular aperture of 60 degrees, the X-ray burst would reach out 190 light-years. However, mankind could fire off this X-ray burst only once!

So X rays come off just as badly as optical lasers. And yet the satellite detectors of the high-energy astronomers are routinely analysing phenomena which change over an interval of milliseconds. We may expect that improved X-ray telescopes will soon be able to observe microsecond processes. Will they be able to observe nuclear explosions?

The curious waves of gravity

Let us now turn to gravitational waves, a more unorthodox type of radiation which may one day be harnessed for communication with extraterrestrial life-forms. As long ago as 1916, Albert Einstein predicted gravitational waves as one consequence of his General Theory of Relativity. But only in the last 20 years, as a result of the rise of electronics and low-temperature physics, has a branch of experimental physics developed which merits the name of gravitational-wave astronomy. This is the attempt to record on Earth gravitational waves originating in this and neighbouring galaxies. According to the theory, any matter which is in any kind of motion emits gravitational waves; in principle even the tulips in the garden when they unfurl their petals under the rising sun. But the gravitational waves from the tulip leaves would be unmeasurably weak. Only if huge masses are set in rapid motion can waves of sufficient intensity be expected to be accessible to measurement with the instruments of the 1980s. To anticipate: man will probably never command a technology capable of setting in motion masses so large that the gravitational waves thus produced will be usable for purposes of communication. This capability must be left to the technically more highly developed civilizations of Kardashev's Types II and III. And even for civilizations which know how to manipulate planets and stars, it seems to me almost nonsensical that this skill should actually be used to operate a transmitter for gravitational waves. In what follows (an extended abstract of ref. 49), I will briefly substantiate this view by describing the principles and the astrophysics of gravitational waves.

Astrophysicists expect gravitational waves to come from distant events, perhaps the most energetic processes possible in the Universe since the Big Bang: supernovae and stellar collapses which produce what we know as black holes, or collisions between black holes themselves.

Since even large amounts of matter are practically transparent to gravitational waves, their sources may be very far away. With new experimental measuring equipment, it is expected that within perhaps 10 years—assuming funding to be available—it will be possible to detect gravitational waves from neighbouring galaxies within a radius of 10 million light-years (Table 11). This is a thousand times the distance from the Earth to the galactic centre.

But how are the gravitational-wave receivers on the Earth to establish changes in distant gravitational fields? By Einstein's Special Theory of Relativity, no physical effect ('information') can travel faster than the velocity of light. But this also implies that in a vacuum (in the absence of a material carrier of information) no effect propagates slower than the velocity of light. If the Sun were suddenly to disappear, 8 minutes later the Earth would fly off into space with a velocity of 30 kilometres a second, as though released from a sling.

Of course, the Sun cannot simply 'make off': the conservation of mass and momentum forbids it. But even a change in the shape of the Sun would make the Earth deviate somewhat from its present orbit. That gravitational effects travel with the velocity of light at least admits of the conclusion that there should be a gravitational phenomenon analogous to electromagnetic radiation. Such 'gravitational waves', like electromagnetic waves, must be transverse, in other words they only vibrate and exert force perpendicular to the direction of propagation (Figure 49).

But the two kinds of waves differ in important respects. The most conspicuous is the fact that gravitation is a far weaker interaction than electromagnetism; for example, the electric force between two electrons is 10^{43} times stronger than their mutual gravitation. For this reason an accelerated electron will emit 10^{43} times as much electromagnetic as gravitational radiation. Gravitation would therefore always be vanishingly small if it were not superior to electricity in a decisive respect: gravitational forces can only reinforce one another. In electromagnetism there are two types of charge, which either attract or repel.

Many electric particles are practically neutral in their outgoing effect, since the field of a positive charge is nearly neutralized by that of a nearby negative charge. But there is only one type of gravitational charge, the (positive) masses which

Table 11 Sources of gravitational waves

Source	Energy (watts)	Frequency ν (wavelength λ)	Remarks
Oscillator with 1000-kilogram spheres 0.1 metre apart and 1 centimetre amplitude, 500 hertz	10^{-40}	1000 hertz	Too weak for laboratory detection
Big Bang	?	$\lambda \geqslant$ diameter of galaxy	Energy flux $< 10^{-4}$ watts per cm², $h < 10^{-7}$ (λ per 10^6 light-years) [2]*
Jupiter orbiting Sun (12-year period)	10^3	$\lambda =$ several light-years	
Binary star (2 solar masses, period a few hours)	10^{26}	$\nu = 2$ per period	Too weak if period > 10 days; strongest known source, the binary Iota Boötis, produces 10^{-17} watts at the Earth, $\nu = 7.5$ per day, $h = 6 \times 10^{-21}$
Crab Pulsar (rotating neutron star of period 0.0333 seconds)	10^{31}	60 hertz	Energy flux 10^{-20} watts per cm², $h < 10^{-27}$; other pulsars are weaker by a factor of 400 or more
Large black holes (10^5 to 10^8 solar masses)	10^{32}	10^{-1} to 10^{-4} hertz	Probably exist at centre of our Galaxy; every time a star of solar mass falls into one, we receive 10^{-10} watts per cm², $h \sim 10^{-19}$
Young neutron star	10^{38}	10^3 hertz	Formed shortly after collapse; energy flux $\sim 10^{-6}$ watts per cm². Galaxy: $h = 10^{-22}$
Explosions in quasars and nuclei of galaxies	10^{39}	Periods ~ 100 days to 3 hours	Gigantic explosions, strong radio sources $h \sim 10^{-21}$; energy flux $\sim 10^{-19}$ watts per cm², much weaker than from binaries
Supernovae, stellar collapse	10^{44}	10^2 to 10^4 hertz	Occur several times per century in our Galaxy; energy flux 1 to 10^3 watts per cm². Virgo cluster: at least once per month; energy flux 10^{-6} to 10^{-3} watts per cm²; $h \sim 10^{-22}$ to 10^{-20}

* h denotes the dimensionless amplitude of gravitational waves (for the Sun it would correspond to $h \sim 10^{-6}$).

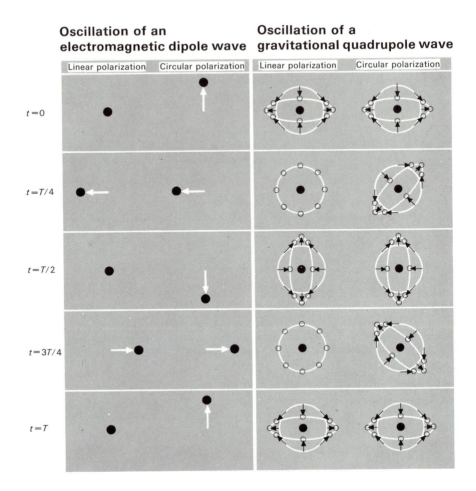

Figure 49 Under the influence of an electromagnetic wave, an electric charge oscillates relative to a neutral particle (left); under the influence of a gravitational wave, a test particle oscillates relative to a central particle (right). The electromagnetic wave is a dipole wave, the gravitational wave a quadrupole wave. In each case, the initial state (at time $t = 0$) and the four phases of a complete oscillation (T = period of oscillation) are shown, for both linear and circular polarization.

always attract one another. As a result, no gravitationally neutral bodies can be made: for this we should need particles with 'negative' mass, which has not yet been discovered.

A third property distinguishes the gravitational field from the electrical field. According to legend, Galileo in 1590 dropped bodies of different weights from the (even then) Leaning Tower of Pisa, and found that they all fell at the same rate, whether light or heavy. For 2000 years, until then, people had believed Aristotle's conclusion that bodies of different weights fell with different speeds.

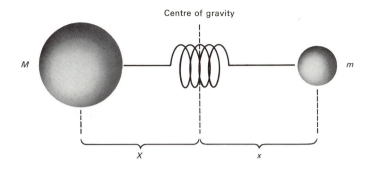

Figure 50 Simplest version of a receiver for gravitational waves: two masses connected by a spring.

For electric charges, on the other hand, the acceleration in an electric field depends on the mass: different electric charges 'fall' at different speeds. This difference has far-reaching consequences for the radiation and detection of the two types of waves. A transmitter or receiver for electromagnetic waves can consist of a charge moved to and fro by a spring attached to a neutral body.

But for gravitation there are no neutral bodies. This means that not only the mass of an isolated body remains constant, but also the position of its centre of gravity (law of conservation of momentum: action + reaction = 0). So a simple source or receiver of gravitational waves must consist of two masses which are moved towards and away from one another by a common spring (Figure 50).

Gravitational radiation thus cannot be directly observed, since it accelerates all particles in the same way. For example, we do not feel the pull of the Moon directly, but only observe it indirectly through the tides, which result from the decrease of the pull over the diameter of the Earth.

Nearly all matter is practically transparent to gravitational waves: they pass through the Earth almost without 'noticing' it. They are even harder to observe than neutrinos. Thus, gravitational waves can bring information from far distant parts of the Universe, which would be invisible to us by other means. The construction of gravitational aerials may therefore perhaps open up to us the totally new field of 'gravitational astronomy', which may prove itself just as fruitful as radioastronomy has been during the last three decades.

What sources can we suggest as emitters of gravitational waves? The gravitational radiation of a laboratory transmitter would be much too weak: two spheres each of 1000 kilograms, 1 metre apart, vibrating relative to one another a thousand times a second with an amplitude of 1 centimetre, emit about 10^{-40} watts in gravitational waves, an unmeasurably small output.

It is equally hopeless to attempt to measure even the gravitational radiation sent out by large nuclear explosions. All conceivable terrestrial sources of this kind of radiation are far too small for any aerial hitherto developed.

The radiation output is the greater, the more mass is contained in the transmitter. Jupiter, orbiting the Sun with 318 times the Earth's mass, radiates gravitational waves with an output of 1 kilowatt. We expect more radiation from massive, close double stars or from rotating neutron stars, the pulsars.

At least some, if not all, supernovae leave behind a rotating neutron star. Some per cent of the total energy Mc^2 of a rapidly-rotating neutron star of mass M (c is the velocity of light) is contained in its gravitational binding energy.

A large part of this energy is probably radiated as gravitational waves at the collapse of the star, when matter is compressed into a neutron star at the centre of the supernova. In the first few seconds after the explosion, the neutron star can emit several pulses of gravitational radiation of about 10^{44} watts with a frequency of 1000 to 10 000 hertz, probably within a few milliseconds.

If more than 2 solar masses are compressed at the centre during the supernova explosion, gravitation will overcome the outward pressure of the neutrons, and the collapse will continue: within thousandths of a second a black hole is born. Here it is likely that as much energy is produced as at the birth of a neutron star (see Part I: 'When the Sun dies . . .', page 24).

More than two dozen research groups are today attempting to construct receivers for gravitational waves. Whether they are aluminium cylinders at low temperature, or quartz crystals weighing several pounds, or laser devices (see Figure 51), the sensitivity of all the equipment, the 'aerials', is still too low. And these aerials for gravitational waves are growing more and more costly. But perhaps in a few decades gravitational-wave astronomy will be as much one of the normal tools of astronomers as radioastronomy and X-ray astronomy are today.

Interstellar communication by the neutrino channel?

Alongside gravitational waves, a further potential, though as yet remote, possibility for the transfer of information is offered by the elementary particles known as neutrinos. Neutrinos are particles without mass; they can thus move through space as fast as light or radio waves. As long ago as the 1930s, the theoretician Wolfgang Pauli had postulated the existence of such particles, but they were not observed by experiment until 1956, with the nuclear reactor in Savannah River, Georgia.

Neutrinos—'the little neutral ones'—fully live up to their name: almost the entire Universe is 'transparent' to them. Neutrinos can even penetrate un-hindered through a lead wall as thick as the distance to the nearest star, Alpha Centauri (4.3 light-years). For this reason, it would not be difficult to send a beam of neutrinos through the middle of the Earth to the other side. A human being directly hit by a powerful ray of neutrinos would notice nothing.

The difficulty, however, consists in collecting these strange particles. In practice, it has only been possible to detect a very small fraction of all neutrinos

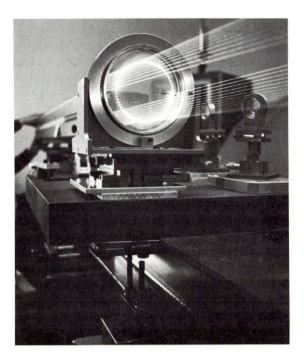

Figure 51 Researchers at the Max Planck Institute for Physics and Astronomy in Munich propose to use this laser apparatus as a model to establish whether it may be possible to capture gravitational waves with an aerial of this type. For this purpose a laser beam is repeatedly reflected back and forth between two mirrors (note points of light on mirror), before being superimposed on a second beam. In order to detect the extremely weak gravitational waves even from supernova explosions in neighbouring galaxies, the mirrors must be separated by several hundred metres. (Courtesy: Blachian.)

considered to exist. Their weak coupling to matter is both an advantage and a disadvantage of the new field of 'neutrino astronomy'. Neutrinos, produced in the deep interior of stars by nuclear fusion, are enabled through their low interaction with the rest of matter to break out of their stars into surrounding space. It is only with a neutrino telescope that it is possible to look into the burning interior of a star.

Exactly this has been the aim of Raymond Davis Jr of the Brookhaven National Laboratory for some years. In the depths of a South Dakota gold mine he has set up a 400 000-litre tank filled with the chemical detergent perchlorethylene (C_2Cl_4)—a fluid which is particularly sensitive to high-energy neutrinos. The essential purpose of the experiment is to detect the neutrinos from the interior of the Sun; but his results have not pleased astrophysicists. Much to their discomfort, Davis has 'spotted' only a third of the neutrinos predicted on the basis of theoretical models of the constitution of the Sun. (On average, Davis picks up one neutrino every 6 days!)

In the meantime, plans have been drawn up to take over whole areas of the Pacific Ocean to produce a detector for solar neutrinos. For Project DUMAND (Deep Underwater Moon and Neutrino Detection), a region of the ocean floor near Hawaii, several square kilometres in area, is to be sown with acoustic and optical detectors to measure neutrinos.

But what is the significance of all this for an interstellar communication network? Mankind is already fairly active as an *emitter* of neutrinos: every 8 minutes, the proton accelerator at the Fermi Laboratory in Batavia, Illinois goes into action, and does more than impart massive energies (400 giga-electron volts) to the protons; at the same time, it also produces, for a split second, a pulse of 10 000 million (10^{10}) neutrinos. These neutrino pulses are very tightly beamed, with an angular size of only 0.1 of a degree. But in order to register even a single one of these neutrinos in each pulse at a distance of 1 kilometre, physicists would already need a bubble chamber made from about 1000 tonnes of steel.

At astronomical distances, therefore, neutrinos can only be picked up if detectors of enormous mass are set up. Suppose that some one at Tau Ceti, a nearby star only 12 light-years away, wished to send us a message with neutrinos produced by proton pulses of twice the energy available in the Fermi Laboratory: to collect these neutrinos on the Earth, our detector would have to have 10 times the Earth's mass.[50]

Thus, the outlook for interstellar neutrino communication is poor in the extreme. And even unrealistic rates of increase in the intensity and duration of the beam do little to change this. Let us imagine that it was possible to increase the duration of the short neutron pulses by a factor of 5000 and raise the energy of the protons by all of 20 times. Even in these conditions, the corresponding receiver—tuned to the Tau Ceti neutrino channel—would require a volume of 1 million cubic kilometres: this would be equivalent to a cube 100 kilometres on each side.

We should not, however, be grudging about these imaginative speculations. For safety's sake, Davis's neutrino data were scrutinized for signs of an artificial message—but with a negative result. 'Technologically the transmission of strong modulated neutrino signals would be a sure sign of an advanced civilization.'[50] It can do no harm, therefore, if we keep a search for them, even with our modest equipment. It is possible that neutrino pulses may be ideally suited as call signals for interstellar messages. As soon as this had led to contact being established, it would be simple enough to change over to the more convenient electromagnetic waves.

Tachyons: particles faster than light

In science it has often happened that things considered theoretically *possible* were *actually found* by experiment. Many of the elementary particles, such as the neutrino or the pi meson, were first predicted on the basis of theory and later found experimentally. With tachyons we are dealing with the concept of particles with 'imaginary' mass (nobody knows as yet just what that is supposed to mean), which never travel with a velocity *less* than that of light. The theory by which such particles are at least not impossible is an extension of the Special Relativity Theory developed by Einstein in 1905. Just as Einstein's original theory describes a mechanics of material particles which never travel faster than light—only light, or particles of rest mass zero, travel with exactly the velocity of light—the Special Theory of Relativity can in principle be reformulated for particles of another kind.[51] This can be done if we introduce particles with another kind of mass—the so-called tachyons. Tachyons have no upper limit to their velocity, only a lower one, and their motion is thus quite peculiar: if we supply them with energy they travel more slowly.

The physics of tachyons, for example how they react when they encounter particles of normal matter, is not yet fully developed, even theoretically. Though many physicists have been concerned with tachyons in recent years it also remains undecided whether tachyons should not be 'forbidden' if they were to contravene the fundamental causality assumptions of physics. According to these postulates the cause always precedes the effect. Physicists usually designate theories as 'unphysical' if a time paradox can be built on them: what happens if I go back into the past and destroy my own ancestors? The tachyons, travelling faster than light, might be able to travel back into the past and perhaps remove the causes for our existence.

In the framework of such a theory, the future would no longer be distinguishable from the past; 'before' and 'after' could no longer be kept apart. And this could lead to a situation which, in a tachyon-less world, could only apply to supernatural beings: *everything* lies in the future, even the past.

In order to demonstrate tachyons by laboratory experiment, an attempt has been made to employ an effect which also occurs in the behaviour of charged particles such as electrons and protons. When an electric charge moves through a medium such as a gas, it emits so-called Čerenkov radiation if the charge races through the medium 'too fast'. Čerenkov radiations are electromagnetic shock waves, analogous to the acoustic shock wave in a Mach cone, which accompanies a supersonic aircraft in the medium of air. The idea was that (charged) tachyons must also emit Čerenkov waves, since they move faster than the velocity of light in a vacuum, even when they are shooting through an electromagnetic field free of matter. It was even calculated what shock waves tachyons would emit as gravitational waves.[52] But up to now, all experimental attempts to detect tachyons in the laboratory by the Čerenkov effect have come to nothing.

In the mean time it was suggested in 1975 that tachyons are surrounded by a field with a special structure—that it is quite impossible for them to emit Čerenkov radiation;[53] thus these experiments were foredoomed to failure. Australian physicists at the University of Adelaide on the other hand attempted in 1977 to detect tachyons in cosmic rays. When high-velocity particles from the Cosmos impinge on the atmosphere they collide with molecules, and a whole shower of other ('secondary') particles is usually produced. If there are tachyons among them, they will outstrip the particle shower and arrive at the detector on the ground before the other particles. It was even asserted at the time that such events had been observed more often than would be permitted by chance. But since then we have heard no more of the Australian experiment, so that for the present it should not be taken seriously.

Be that as it may, if tachyons did exist, their unlimited speeds would make them favoured bearers of information over great cosmic distances. This assumes, of course, that the appropriate tachyon telescopes and tachyon receivers can be constructed—a relatively minor detail in the context of this essentially speculative discussion! It is still of interest, however, how many bits per unit time could be transmitted with the aid of tachyons.

The amount of information which can be transferred reflects the strange physics of tachyons. The faster the tachyons, the fewer bits can be transmitted. Conversely, for 'slower' tachyons—slower than 10 000 times the velocity of light—several orders of magnitude more bits can be carried per second, and with less expenditure of energy, than by light. Only at 100 000 times the velocity of light and upward do the bit rate and the energy expenditure become comparable with communication by light. At the same time, these tachyons would make possible an exchange of information throughout the whole Universe in less than 100 years. The quickest exchanges would be of messages with the smallest information content. At the (theoretical) infinite speed possible for tachyons only silence would remain.

Physicists have as yet little cause to take tachyons seriously. A possibility which appears equally exotic, if entirely realistic, is the exchange of information by the 'biological channel'—biological communication.

Biological communication

Alongside the various means of communicating with extraterrestrial intelligences that we have already discussed, a further interesting possibility has clearly been overlooked until now: the biological communication channel. It was recently suggested by two Japanese researchers, Hiromitsu Yokoo and Tairo Oshima, that extraterrestrial messages might be concealed in the genetic material of certain viruses—those known as bacteriophages—or bacteria themselves.[54] Although terrestrial biochemistry is not yet in a position to take such a step, the rapid advances being made in genetic engineering nevertheless make it plausible that other, older civilizations may be using this process.

There is no doubt that this communications technique offers certain advantages compared with other methods, such as spaceships or radio signals. Contact by spaceship, though direct, costs a great deal and is slow in producing results. Radio contact, although equally direct, as well as being cheap and as fast as natural laws allow (that is, as fast as light), is less reliable, as the sender can never be certain whether his signals will be received. The information transmitted is also destroyed by background noise and interstellar dispersion as the distance increases.

The advantages of biomolecular contact are these:

First, it would be a simple matter, even for our present civilization, to send the molecules into space, enclosed in capsules for safety, either in all directions or aimed at particular stars.

Once the microorganism containing the message has landed on a suitable planet, it can reproduce. The problem of noise does not arise, as the biological message can be copied along with the rest of the organism and spread throughout the entire planet. Reproduction errors would be excluded, as mutations will die out if the microorganism is already optimally suited genetically to its environment. For the same reason, biocommunication, compared to radio signals, is easy to receive: the messages multiply themselves and remain on hand after delivery for as long as the organism retains its ecological niche—until an intelligent life-form finally decodes the message.

The greatest disadvantage of this method, in the view of the Japanese researchers, lies in the fact that the messages can only be sent to planets whose biochemistry is identical to that of the sender; moreover, viruses must be sure of finding a suitable host bacterium in order to procreate.

If the sender was confident that the hereditary substance (DNA) of the smallest viruses which attack the commoner bacteria would be known to the receiver, the message would stand a fair chance of being received. It is now 2 years since biochemists established the full sequence of the 5375 genetic 'words' of the bacteriophage classified as ΦX174, which is one of several hundred viruses which attack coliform bacteria. Its DNA 'alphabet' consists of four 'letters' (the nucleotides); each word ('codon') consists of three letters—the nucleotide triplets discussed in Part I (page 41)—and the words join together to form 'sentenses'—the genes. Each gene is responsible for the synthesis of a particular protein (see Part I: 'The primeval soup and evolutionary reactors', page 38). In the bacteriophage ΦX174, the genes have the remarkable property of 'overlapping'; and just as, when two sentences are strung together without punctuation, some words may make sense considered as part of either, sections of a gene also belong to another gene.

If any message is concealed anywhere in these genes, it will be in these bridging words—especially as, to quote Yokoo and Oshima, 'At present it is quite difficult to explain the origin and evolution of overlapping genes in terms of molecular evolution.'[54] There are three locations in ΦX174 where genes overlap, consisting of 121, 91 and 533 words respectively. And this is where an already odd phenomenon turns up a further surprise: each of these counts is the product of two prime numbers—$121 = 11 \times 11$, $91 = 7 \times 13$ and $533 = 13 \times 41$.

This gave the Japanese team the idea of reading these particular sequences of words as cosmic pictograms, drawing them up in the style of the Arecibo cosmogram, with the lengths of the sides given by the prime factors, for instance as a square of 11 rows and 11 columns. First they gave each of the four letters of the DNA alphabet a different colour, and looked for intelligent patterns—without success. Then they picked out the first, second, or third letter in each word separately. In one version, they represented the letters A and G (the purines adenine and guanine) as black squares and T and C (the pyramidines thymine and cytosine) as white squares. As a variant on this, they tried representing the complementary pairs C and G in black and A and T in white (see Figure 52). However, none of the various attempts produced any hint of an encoded message.

Let us now turn to signals which bear some of the features of an artificial source of radiation. The fact that a 'natural explanation' has so far always been forthcoming should caution us against accepting claims of artificial origin.

A hole at the centre of the Galaxy

Many unusual signals reach the Earth, and these have often led otherwise objective scientists to put forward hasty and adventurous interpretations. The reason for this lies in the unexpected discoveries with which astronomy regularly surprises us. Since the discovery of such exotic objects as quasars, pulsars and X-ray stars we have learned that many new types of sources can also display properties otherwise attributed to artificial sources. The extremely small angular size which we made a criterion for artificial signals (see 'Artificial message or natural noise?', page 99) is also characteristic of cosmic objects discovered in 1976 and 1977: gigantic black holes in the centres of galaxies.

Powerful X rays were observed in 1976 in the galaxy Centaurus A, apparently emanating from a very small object. A group of Cambridge astrophysicists published the explanation that a very large black hole—perhaps with a mass 10 million (10^7) times that of the Sun—might be the source of these X rays.[55] Centaurus A has long been known as an elliptical galaxy whose radio waves originate at two points exactly symmetrically placed relative to the centre of that galaxy. It has been suspected for some time that this symmetry is no accident, but that a common central energy source causes this symmetrical radio emission. This central body of Centaurus A emits hard X rays from a volume only 10 times the size of the Solar System.

Astrophysicists believe that large amounts of radiation were liberated long ago by energy-rich processes between the stars in the central star clusters of the

Figure 52 Japanese scientists have attempted to decode an extraterrestrial message they suspected of being concealed in the sequence of genetic 'words' of the bacteriophage ΦX174. The 121 overlapping words belonging to both gene A and gene B have been arranged in an 11 × 11 square on the lines of a Drake pictogram. Each word consists of three of the four 'letters' A, G, T and C, nucleotides of the genetic 'alphabet'. For the top three samples, the first, second and third letters of each of the 11 × 11 triplets was separately picked out, with the purines A and G shown as black squares and the pyrimidines T and C as white squares. As a second variation, the nucleotides complementary to each other in the replication process were paired, with G and C shown in black and A and T in white.

GAA	TGG	AAC	AAC	TCA	CTA	AAA	ACC	AAG	CTG	TCG
CTA	CTT	CCC	AAG	AAG	CTG	TTC	AGA	ATC	AGA	ATG
AGC	CGC	AAC	TTC	GGG	ATG	AAA	ATG	CTC	ACA	ATG
ACA	AAT	CTG	TCC	ACG	GAG	TGC	TTA	ATC	CAA	CTT
ACC	AAG	CTG	GGT	TAC	GAC	GCG	ACG	CCG	TTC	AAC
CAG	ATA	TTG	AAG	CAG	AAC	CCA	AAA	AGA	GAG	ATG
AGA	TGG	AGG	CTG	GGA	AAA	GTT	ACT	GTA	GCC	GAC
GTT	TTG	GCG	GCG	CAA	CCT	GTG	ACG	ACA	AAT	CTG
CTC	AAA	TTT	ATG	CGC	GCT	TCG	ATA	AAA	ATG	ATT
GGC	GTA	TCC	AAC	CTG	CAG	AGT	TTT	ATC	GCT	TCC
ATG	ACG	CAG	AAG	TTA	ACA	CTT	TCG	GAT	ATT	TCT

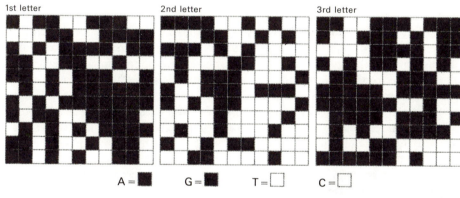

1st letter   2nd letter   3rd letter

A = ■   G = ■   T = □   C = □

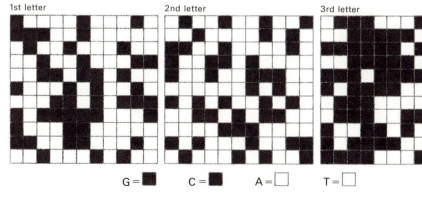

1st letter   2nd letter   3rd letter

G = ■   C = ■   A = □   T = □

galaxy—processes still not understood. These spread along the rotation axis of the galaxy towards its edge, symmetrically in both directions. From two 'bubbles', the radiation then reaches us in the form of radio waves. The remnants of the nuclear burning of perhaps 10 million suns remained at the centre. From this compacted mass of material, 20 times the size of our Sun, a black hole could have formed. Matter in the neighbourhood of the star cluster, falling into the black hole and thereby heating up to several hundred million degrees, could be converting the energy of its fall into X rays such as we observe.

The existence of black holes, however, must always be deduced from radiation processes in their immediate vicinity; and the 'proof' is similarly indirect. But as long ago as 1971 the British astrophysicists Donald Lynden-Bell and Martin Rees nevertheless suggested that, in searching for large black holes at the centres of galaxies, a look-out should be kept for sharply localized radiation phenomena of this sort—not only X rays, but infrared, optical and radio emissions. Thus, when a minute source of radio waves was detected in June 1977 at the centre of our Galaxy (that is, 30 000 light-years away), the discoverers—a team of American radioastronomers at the National Radio Astronomy Observatory in West Virginia (where the Ozma I Project was initiated in 1960)—supposed the case to be similar to that of Centaurus A.

According to their estimates, the object there could be a black hole with a solar mass of up to 10^7. Such enormous masses are by no means unusual for the central region of a disc-shaped spiral galaxy like ours. But at the same time a black hole of this mass occupies an extremely small volume: it would take up about 20 times the distance from Earth to Sun. By way of comparison, Pluto, the outermost planet of the Solar System, orbits the Sun at 40 times the Earth's distance. What is striking about the American measurement is that the resolving power was great enough to detect so tiny an object emitting radio waves at a distance as great as that of the galactic centre. This was made possible by Very Long Baseline Interferometry (see 'Artificial message or natural noise?', page 99).

In their investigation of the Milky Way the researchers used two telescopes on the East Coast of America and one in California, separated by about 4000 kilometres. The data indicated an extremely compact radio source in the constellation Sagittarius (the direction in which the centre of the Galaxy lies), a hundredth of a second of arc in diameter with a powerfully radiating nuclear region a thousandth of a second of arc across.[57] Tiny as this radio source was, none of the astronomers supposed that it might be of artificial origin. On the contrary, the properties of the radiation indicated rather that—as with Centaurus A—it is produced by very fast-moving matter as, for example, when falling into a black hole under the influence of its gravity. These huge black holes could have originated very early, in the primitive beginnings of galaxy formation.

Up to now, the new object in the centre of the Galaxy has shown no trace of localized X-ray emission like that of Centaurus A. But our theoretical understanding of radiation processes in the vicinity of black holes swallowing up matter is not yet well enough developed to clarify such differences in every detail. And it is all the harder to draw reliable conclusions concerning the existence of a technological civilization when we are dealing with the strange radiations which reach us from the depths of the Cosmos. Until we directly intercept artificial signals, such a hypothesis can only be 'proved' in as indirect a way as the existence of enormous black holes in the hearts of galaxies.

Quasars—galaxies the size of stars?

They radiate as much as a whole galaxy, yet appear no larger than stars. Their light is more strongly red-shifted than that of any other celestial body. But in visible light they are unremarkable objects, to which astronomers paid no particular attention until, in the mid-1960s, radio telescopes detected them as very vigorous point sources—an uncommon combination for radio sources. For this reason they were called *quasars*, quasistellar objects (QSOs). Only then was their visible light examined, for example with the Mount Palomar telescope. But it was by no means easy to decide which of the many faint stars on the photographic plates matched a strong radio source, for the positions of radio sources cannot be determined with much precision with radio telescopes. So it was some time before (in 1962) the Australian astronomers identified a star-like feature named 3C 273 (No. 273 in the third star catalogue of the University of Cambridge) as the optical counterpart of the previously known radio source (Figure 53). This was done with the aid of the Moon. When the Moon passes over the source, and thus eclipses it, the position of the source can be precisely determined from an accurate knowledge of the Moon's orbit and the time of disappearance. From this it was established which of the many stars in the neighbourhood of the uncertain radio position was emitting the radio waves: it was the brightest of the stars in question, in fact, the brightest optical source of all quasars hitherto detected. It was recognized only in the following year that the quasar 3C 273 was a quite peculiar type of star. The Californian astronomer Maarten Schmidt photographed the spectrum, but the broad, bright spectral lines did not seem to fit the patterns of any of the known atoms (such as hydrogen). On 5 February 1963, Schmidt hit on a plausible explanation—the lines fitted hydrogen if it had undergone an unusually marked redshift (Figure 54).[58] Schmidt recalls: 'That night I went home in a state of doubt. I said to my wife, "Today something really incredible happened to me."' The redshift indicated that 3C 273 was receding from us with more than a tenth of the velocity of light. If this motion relative to the Earth is caused by the expansion of the Universe, quasars are the most distant objects known to us. Although still controversial, this 'cosmological' interpretation of the redshift is no longer seriously doubted by many astrophysicists. In the optical region, quasars, because of their great distance from us, resemble faint stars.

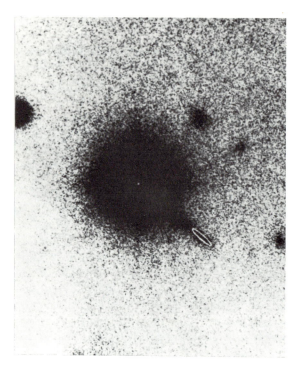

Figure 53 This first and optically brightest quasar, astronomically designated 3C 273, was discovered in 1963. The radio emission of this quasar has two components, indicated by the white dot at the centre of the object and the oval on one of the ray-like arms. Perhaps these most distant objects in our Universe are gigantic black holes at the centres of young galaxies. (Courtesy: Royal Astronomical Society, London, RAS 682.)

More than 600 quasars are known so far, with recessional velocities up to over 90 per cent of the velocity of light, or over 270 000 kilometres a second. Some are up to 10 000 million (10^{10}) light-years from us. The widths of the spectral lines of quasars show that matter must be moving with a velocity of thousands of kilometres a second within the quasars too. The brightness of many quasars varies, often irregularly for years (as for 3C 273), but sometimes periodically. The quasar 3C 273 is also emitting streams of high-velocity gas,[59] a phenomenon which has also been noted for other quasars. Further glimpses into the secrets of the quasars were gained in 1977, when astronomers at the Kitt Peak Observatory in America observed that two sources which did not emit observable radio waves, and hence had been regarded as variable stars, were actually quasars.[60]

What makes quasars so mysterious—even today they are little understood—is the intensity of their light. The quasar 3C 273 is five million million (5×10^{12}) times as bright as the Sun—the radiation of an average galaxy concentrated in a region only a few light-years in size. This compactness of quasars is a property we would expect in artificial sources. Are the astroengineers of a supercivilization at work here?

At present this hypothesis is by no means so easy to refute, so long as scientists

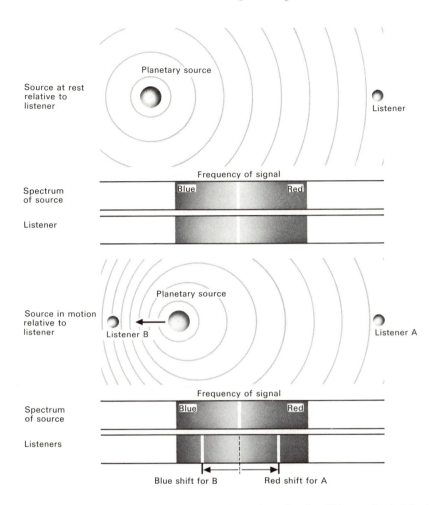

Figure 54 Doppler effect: if the source is in motion, signals will be received shifted in frequency.

have not advanced a convincing 'natural' explanation of the quasar phenomenon. However, it must be said that further striking properties of quasars, besides their smallness, also suggestive of intelligent activity, would have to be found before we were bound to undertake a more serious analysis of this hypothesis. However that may be, Very Long Baseline Interferometry, with its capacity for high resolution, will provide us with further insights. With its aid we are already seeing into the very 'interior' of the quasars!

These radio observations of quasars currently suggest that:

• The radio waves are produced by electrons in spiral motion in magnetic fields at velocities close to that of light.

• These electrons are localized in clouds only a few light-years in extent—some of the most active quasars even emit fresh clouds from time to time.

• Quasars and the central regions of active galaxies display great similarities: perhaps quasars are the centres of young galaxies in process of formation.

Theoreticians currently favour several models for quasars and the active nuclei of galaxies; two theories lead the field at present. The first maintains that we are dealing here with enormous black holes, which are sucking in stars and gas clouds. This model is supported by the astronomers in Cambridge, Pasadena (California) and Munich.[61] The second theory introduces the concept of *magnetoids*. These are supermassive giant stars with strong magnetic fields (similar in shape to the magnetic field of the Earth, but much more powerful), rotating like a gigantic bar magnet. This model is still particularly supported by some Moscow astrophysicists,[62] since the rotating magnetic field of magnetoids would explain the nearly periodic variations in brightness shown by some quasars.

It is also interesting to consider what happens to the quasars after they 'die', for their activity usually ceases after around 10 or 100 million (10^7–10^8) years. Martin Rees in Cambridge estimates that there are today perhaps 100 000 times more 'dead' than 'live' quasars.[63] Rees, who has devoted himself completely to the quasar model involving huge black holes, thinks that there is a now silent quasar in almost all large galaxies. Perhaps the compact radio object in the centre of our Galaxy (described in the previous section) was once a quasar, if never a very exciting one.

Stars faster than light?

Not only the details of quasars, galactic nuclei and massive black holes can be examined with the techniques of Very Long Baseline Interferometry. This technique has also uncovered another detail, previously unknown, of certain compact radio sources outside our Galaxy. Several of these objects often consist in reality of two components—they are double sources, receding from one another.[64] This discovery has a spectacular consequence: if the radio sources are as far away as their redshift indicates, it seems that they are separating with velocities several times that of light (at right angles to our line of sight). This discovery naturally stimulated the brains of a number of speculative physicists, especially those who were in any case inclined to consign otherwise well-established theories to the waste-basket at the drop of a hat. Are we seeing particles moving in some cases at 10 times the speed of light? And who (or what sort of mechanism) has accelerated them to this speed? Is it possible that we have (at last) found extraterrestrial beings at work? Are they applying laws of nature we are still ignorant of?

It is true that a scientist loves to be stimulated, but when he is, it is (usually) with some method. First we must realize that phenomena which *appear* to exceed the velocity of light are perfectly conceivable and in principle possible; but these do not require any energy (or mass) to be transported at a velocity greater than that of light (which, according to the Special Theory of Relativity, cannot happen). For example, imagine two very long, thin rods laid almost parallel and close together, so that they just overlap at one end. Now, if we can only move one of the rods fast enough across the other, it will appear to the observer that the *point of intersection* is moving away from him faster than the speed of light. But in fact no law of physics is contravened, as each rod is really moving at less than light speed. We are only seeing an effect of projection.

Besides this, the postulate that the velocity of light cannot be exceeded is fundamental to modern physics, and countless predictions based on it have been verified. Therefore the astronomers first proceeded to re-examine the data. Measurements with Very Long Baseline Interferometry are in fact very difficult to interpret—the interpretation must always be built on some assumption, for example the assumption that two radio components are moving apart in the observed section of the sky. But if in fact a tiny third radio source appears near the two others, analysis of the data can easily create the impression that the two light sources are separating at more than light speed. Therefore the original observations have been repeated at intervals during the past 10 years, with the result that they unequivocally confirmed the first interpretation, so that at least this doubt can be regarded as removed. The other explanation suggested, that the large distances derived from the redshifts may be incorrect, is rejected by most astronomers. (The reason is that when we try to explain the redshift by anything other than the recessional velocity we jump right out of the frying pan into the fire.)

Nobody seriously believes that physical laws are being broken by this phenomenon. However, it must be granted, on the other hand, that there is still no astrophysical model which will satisfactorily accommodate all the known facts about these faster-than-light radio sources. Although the data permit of no final conclusion, there is nevertheless every indication that some kind of stream of particles is striking a sort of 'screen' at near light speed, and emitting the observed radio waves as a result of scattering by the screen. If the particles fall at the right angle on a screen of appropriate shape, the desired projection effect can be produced.

The rapidly expanding double radio sources are typically only a few light-years apart (an indication that they belong to a single source of energy, possibly the active nucleus of a galaxy). The following scenario now seems plausible. A gigantic explosion in the nucleus of a galaxy ejects a shock wave of particles. The shock wave moves preferentially perpendicular to the galaxy, both upwards and downwards, as in those directions it meets the least resistance. Low-density gas outside the galaxy, forming an almost cylindrical 'tunnel' about its rotational axis, can act as a screen. If the shock wave now strikes the walls of this tunnel slantwise above and below, radio radiation is produced at two points above and beneath the plane of the galaxy. Seen from certain angles, this process gives the impression that the two radio sources observed above and below the galaxy are separating with a velocity exceeding that of light.

Unrealizable contacts: Rama, Solaris and The Invincible

Science fiction writers, like scientists, have speculated a good deal on different kinds of contacts with extraterrestrials. Typically we often hear that if contact is made with an extraterrestrial civilization, 'communication will be difficult' but we will 'manage it' somehow.

It has been given to only a few writers to draw a careful picture, at least on a literary plane, of *how* different the life-forms of other galactic civilizations can be—so alien that, despite intelligent approaches, any real contact would be impossible and indeed pointless. The Polish master of science fiction, Stanisław Lem, has written at least three such books: *The Invincible* (1964), *Solaris* (1961) and *Eden* (1959).[65] Here *The Invincible* is of especial significance, as well as an equally fascinating novel, *Rendezvous with Rama* (1973), by the British–American author Arthur C. Clarke.[66] Both authors, like Isaac Asimov, are accomplished scientists, and write directly on scientific subjects. Their science fiction work therefore incorporates contemporary scientific knowledge.

The books in question deal with a favourite theme of science fiction: contact with an extraterrestrial civilization. In *Rendezvous with Rama*, Clarke takes us to the year 2130, in which the Moon and several planets of the Solar System have been colonized. Although certain social customs, such as marriage, have changed, life on Earth is still recognizable as a further development of our society. Suddenly an unidentified object enters the Solar System. It might perhaps be a comet, but then it abruptly changes its course, accelerates and adopts a trajectory which may endanger the Earth. Only one of the Earth's spaceships then on their way across the Solar System is near enough to cross the calculated new path of the object. The crew of the spaceship finally succeed in approaching the flying object; it proves to be a cylinder 50 kilometres long, to which they give the name Rama. They land on its surface and enter its interior, where a world resembling a landscape, inhabited by biological robots, is revealed. To all appearances their task is to take care of the ship during its journey to an unknown goal. The terrestrial spacecrew come upon complexes of windowless buildings in which information about a civilization—apparently that of the constructors of the Rama—is stored. Finally Rama heads for the Sun, draws upon its energy, and swings out of the Solar System on an accelerated course. It leaves mankind perplexed, and none the wiser but for one fact: they are not alone in the Universe; but they have not even learned the whereabouts of Rama's home planet.

The spaceship Invincible, in Lem's novel of that name, is sent on a special mission. Its task is to find why a sister ship, the Condor, has not returned from Regis III, a previously unexplored planet in the constellation Lyra. On this desert-like planet the men and women of the Invincible find—alongside the remains of a perished civilization—the externally uninjured bodies of their predecessors near the Condor. The mystery of their death is at least partly solved when they come on the presumed agents of the catastrophe which destroyed the crew of the Condor: metallic crystals of microscopic size. These crystal mechanisms probably originated with a race of robots which landed on Regis III with a civilization which has since died out. The robots have outlived their constructors. The scientists on the Invincible suspect that the limited supply of raw materials on this planet gave rise to evolution among the robots. After they had exploited the ground resources of Regis III they began to dismantle each other—to the disadvantage of the larger robots, whose energy requirements were larger and who therefore 'reproduced' more slowly. A struggle for survival broke out among the robots. Smaller mechanisms, though 'intellectually' inferior to the large robots, needed less energy, and finally triumphed over their larger brothers.

The crystals swarm in clouds like flies, guided only by their newly developed collective intelligence, united by their own electromagnetic fields. They destroy objects which threaten them by surrounding them with such powerful electro-magnetic fields that all the memory content of the object, such as that in a human brain, is extinguished. This is how the crew of the Condor must have died. Any communication with these robot flies is pointless: their sole capacity for perception now serves only their own collective survival—a degeneration of the original service function of their robot ancestors. The spaceship Invincible resigns itself to leaving this planet—hostile to all organic life. In a last foolhardy rescue attempt, a member of the crew goes in search of strayed companions. Shielding his head with a metal net, he searches for his friends among the robot flies and unseen by them, and first becomes conscious of the radically alien nature of this 'life-form':

> He felt himself so superfluous in this domain of perfected death, where only dead forms could rise victorious to carry out mysterious rites that would never be witnessed by a living creature. . . . He knew that no scientist would be capable of sharing his feelings, but now not only did he desire to return and to report what he had found out about the death of his companions, but also to demand that this planet be let alone. Not everything everywhere was intended for us, he thought, as he climbed slowly down [*The Invincible*, Chapter 11].

A radical alien-ness of quite another kind is described by Lem in *Solaris*, a book which also appeared as a film in the Soviet Union. The entire surface of the planet Solaris is covered by an organic mass, comparable only to a gigantic brain.

The terrestrial scientists, established on this surface with a research station, use every means they can to set up communication with this organism. But the Solaris life-form reacts only sporadically and in an unforeseeable manner on the subconscious of the investigators. Dead wives 'materialize'; the scientists are brought face to face for weeks on end with repressed memories of their personal past in tangible form. After several scientists have met with fatal accidents as a result, they break off the experiment in communication and leave the planet/brain Solaris without any definite results.

Clarke and Lem ('the visionary among the charlatans') have succeeded several times in their stories in predicting technological developments which have been overtaken by reality faster than their contemporaries would ever have thought possible. While scientists in the East and West are analysing all conceivable problems of communication with extraterrestrial civilizations, and examining possible solutions, it may also be useful to keep in mind that extraterrestrials—if we ever meet them—may be more different than anything we commonly dream of.

The astroengineers of supercivilizations

Can we predict the development of a civilization over cosmological intervals of time? In Isaac Asimov's classic trilogy, *Foundation, Foundation and Empire* and *Second Foundation* (1951–3),[67] the protagonist Hari Seldon (11,988–12,069 Great Empire) uses the laws of a branch of science called psychohistory to extrapolate the future destiny of terrestrial civilization as a part of the galactic community. I have no doubt that, by using mathematical laws, it will one day be possible to chart the future development of a civilization more accurately and over a longer time span than is the case today. In certain areas, such as industrialization, energy consumption and destruction of the environment, this is already being attempted—for example, in the analyses of the future by the Club of Rome. But the application of the systematic developments of group dynamics indicated in the comprehensive studies of history by Toynbee and Spengler to the *whole* of mankind, and their extrapolation to the future, still encounters fundamental scientific obstacles. First, there is at present only *one* mankind: it is only reacting with itself; there is no contact, peaceful or otherwise, with other civilizations; we have no means of comparison by which laws concerning the overall development of mankind can be tested. Second, mankind is approaching the limits of its expansion: those imposed by the finite surface of the planet, non-renewable reserves of fuel and restricted foodstuffs. It may only be one or two hundred years before terrestrial energy consumption has a profound effect on the climate. Then significant changes to the Earth's climate can only be avoided by increasing recourse to the direct use of the Sun's energy.

Today the population is increasing more or less exponentially and our knowledge is now doubling every 10 years. Of course, such phases of exponential growth are transitional periods in the development of a civilization, and are very soon brought to an end by geographical and ecological boundary conditions. Only if this growth continues will the economic development of the next 200 years force mankind to expand into the surrounding Universe.

Does a similar destiny await civilizations on other planets? There are stars far older than the Sun. The oldest, however, are poor in heavy metals. Thus, their planets are also likely to have limited metallic surface deposits, which virtually excludes the development of a technological civilization. But conditions are more favourable on planets around stars which are only one or two thousand million years older than the Sun. Civilizations of corresponding age could therefore exist on them. We do not know, of course, whether mankind will survive the next few hundred years, whether further development will not be interrupted by self-destruction or simply by a decline in intelligence. But it is at least conceivable that some other civilizations may have succeeded in overcoming such crises.

Freeman Dyson discussed in 1960 how living creatures in a planetary system like our own might best utilize this planetary system as a place to live in.[27] Dyson qualifies his thesis from the start: 'I do not argue that this is what *will* happen in our system; I only say that this is what *may have* happened in other systems.'[68] His idea is to examine whether ancient civilizations might modify their stellar environment to the extent that such artificially altered solar systems would become astronomically observable. In order to establish the existence of a supercivilization in this manner, it suffices to consider those civilizations which have developed furthest technologically, and from which we can expect the greatest interference with their environment. We can impute practically anything to a supercivilization, so long as it does not contradict any known physical laws. Dyson therefore makes three assumptions:

• Nothing is too large or too crazy to have been done by at least one technological civilization in a million, provided only that it is physically possible.
• All engineering projects are carried out using a technology which can be understood by mankind in 1965 (although this certainly represents an under-estimate of the actual technical capacity of a more advanced society).
• Cost is no object.[69]

In Dyson's view, provided enough time and opportunity are available, everything that is *possible* will eventually be done somewhere by somebody. Dyson also assumes that in principle it is technically possible to use the radiant energy of the Sun and the available planetary material in an optimum manner to provide living space—and that merely with a quantitative rather than qualitative improvement in our technology. For example, we could dismantle the largest—

and otherwise (to us) quite useless—planet Jupiter, in order to construct a sort of spherical shell around the Sun, a little outside the Earth's orbit.

Of course the material of the shell would not form a rigid unit. A solid shell of matter around a star is mechanically impossible. Rather, the biosphere should be imagined as a loose aggregation, a swarm of islands—'artificial' asteroids—which orbit independently at about the same distance from the Sun in paths which avoid collisions. The size of the islands is limited by the carrying capacity of the material used and the gravity at the location of the asteroids. Structures of steel or fibreglass can be self-supporting in orbit around the Earth up to a diameter of 300 kilometres. In orbit about the Sun at the Earth's distance an island could be up to a million kilometres in size without breaking up.

Such large islands could be constructed so as to keep their total mass as small as possible. A suitable form would be a 'hierarchical' structure: 12 steel rods, each 1 metre in length (thickness 1 centimetre) are combined in an octahedron (first generation). A larger octahedron can be constructed from 100 of these smallest structural units by joining surface to surface. This yields the second generation of structural units. From these, as before, octahedra of the third generation are assembled. Four generations are enough for a structure in Earth orbit—300 kilometres in diameter—and for an island in orbit about the Sun, a million kilometres in diameter, six generations of octahedra suffice. The surprising thing is that the higher the hierarchical level of the octahedra, the lighter they become. With increasing size the density of the octahedra decreases. Dyson estimates the minimum mass of 200 000 such islands at one-hundred-thousandths (10^{-5}) of the Earth's mass.

It no longer seems impossible today to dismantle whole planets. More than a cubic kilometre of ore is already demolished every year for industrial purposes. An expansion of the Earth's industries by a factor of 10^{12} would require the processing of a considerable portion of the Earth (whose volume is about 10^{12} cubic kilometres). Industrial plant and raw materials would have to be moved into space. Instead of the Earth, astroengineers can demolish another planet.

Supplies for the inhabitants of Dyson's shell could be provided by a molecular biology developed for the purpose—a technology which Dyson regards as an attainable goal for molecular biology and gene synthesis. Practically all the solar energy would then be absorbed by this loose ring or spherical shell, on whose immense inner surface mankind could extend itself. The shell would have about 600 000 times the Earth's surface. It would probably be a few metres thick, and could be made a comfortable place to live—despite the presence of such equipment (solar cells) as might be needed to exploit the Sun's energy.

Once this solar energy has been used by the inhabitants of the shell, it must in the end be re-radiated outwards. This results in a surface temperature of 200–300 Kelvin on the outside of the shell. In the sky this shell would appear as an object with the diameter of the Earth's orbit, dark in the optical region, but emitting strongly in the 'far' infrared at a wavelength of about 10 micrometres. It happens that infrared radiation between 8 and 12 micrometres penetrates the Earth's atmosphere; these infrared stars would therefore be observable from the Earth's surface.

It should be mentioned in passing that even the expansion of mankind over a shell of matter surrounding the Sun would not remove the necessity of controlling the population explosion. Even a habitable area 600 000 times as large as that available on the Earth would be filled in 1000 years by a 2 per cent annual increase of population. Sebastian von Hoerner points out: 'Even the maximum exploitation of the entire Solar System cannot solve our population problem; it merely postpones it by a thousand years. And after only half of this time, energy would have to be rationed.'[70]

For millions of years the Solar System has kept a mini-version of Dyson's shell—a ring of material—ready for us, in the shape of the asteroid belt, a ring of hundreds of thousands of scraps of matter, some of them several kilometres in size, between Mars and Jupiter. Today it may still seem daring, but some day Dyson's vision may well take on concrete form, beginning with settlement of the asteroid belt. Already two precursors of mankind have passed through the asteroid belt—the American space probes Pioneer 10 and 11, on their way to Jupiter.[71]

What sort of life would be possible in a shell world like Dyson's, in the asteroid belt or perhaps on comets? The gravitational field of such tiny bodies is much too small to retain an atmosphere, and gases artificially supplied simply disperse into the surrounding vacuum. And what would farming be like on a mini-planet which may be much further from the Sun than the Earth? To these problems, too, Dyson has applied his fascinating thinking.[72] Future methods of *biological engineering* and self-reproducing machinery will make life possible even in the less hospitable regions of a solar system. Biological engineering comprises the artificial synthesis of living organsims designed to serve special purposes. Self-reproducing machinery means the imitation of function and reproduction in living organisms by non-living material. With the discovery of the double helix of DNA, the basis of the manipulation of genetic material, of 'genetic engineering', was established. By building on this we can conceive of a biological technology synthesizing artificial living organisms designed for quite specific purposes. With such techniques, if they were ever mastered by mankind, life in space could be made possible. Is this too utopian a picture? There are at least signs that rapid development lies ahead for the science of organic biochemistry. Today we are using microorgansims systematically to produce certain materials, or to destroy pests (biological pest control). Certain bacteria are used industrially in Japan to produce the food additive glutamate. The bacteria in the root hairs of leguminous plants, which supply atmospheric nitrogen to the plants directly, are to be transferred to other plants. (The Society for Research in Molecular Biology at Stockstadt near Braunschweig is making a general study of the commercial possibilities for such processes.) And the cells of mice are stimulated to produce insulin by the introduction of portions of genes (see Part I:'The genetic future of mankind', page 52).

The time may come when we will produce at will microorganisms with suitable enzyme systems, for oil refining in the chemical industry, for food processing and even to combat environmental pollution. Dyson gives the following example of this: 'At present many lakes are being ruined by excessive growth of algae feeding on high levels of nitrogen or phosphorus in the water. The damage could be stopped by an organism that would convert [atomic] nitrogen to molecular [gaseous] form or phosphorus to an insoluble solid. [Or] an organism could be designed to divert the nitrogen and phosphorus into a food chain culminating in some species of palatable fish.'[72] As well as this, the poisonous lead in our rivers and the sulphur oxides in industrial waste gases could be destroyed by suitable organisms and rendered harmless to nature. But in spite of everything this is likely to remain wishful thinking for the foreseeable future. Biochemists I have consulted were very sceptical about the prospects of synthesizing whole genes in order to attain definite, previously determined functions for a cell.

But these examples should serve to direct our thoughts to a vision of how biological engineering could get into gear for the colonization of Dyson shells: on asteroids and on comets. The Sun has thousands of millions of comets, rich in water and in other elements necessary to life such as nitrogen and carbon. What they lack, as the asteroids do, is warmth and air. To Dyson this presents no obstacle to cultivating trees on comets—and this is the point at which biological engineers enter the picture. The problem is to develop plants artificially with leaves accurately designed for the extreme conditions in space: they must be impervious to water, opaque to cell-destroying ultraviolet radiation, able to absorb visible light for the purpose of photosynthesis, and well protected against cold so as not to freeze. Trees with such leaves should even prosper on a comet at the distance of Saturn's orbit.

Finally, further out where the light and warmth of the Sun are weaker still, more complex leaves could permit trees to grow: specialized leaves which would collect sunlight like a concave mirror and deflect and concentrate it on to other sections of the leaf serving photosynthesis. The branches must be ideal heat insulators, the roots might melt the ice within the comets and flourish there, where men live. And how tall could such a tree grow? Comets a few dozen kilometres across have so weak a surface gravity that trees hundreds of kilometres high could grow out into the surrounding space, in order to collect enough energy from the Sun.

In parallel with this, the settlers on the asteroids could supply themselves with the aid of self-reproducing machines. Constructed like living organisms, but of metals and computers, they are able, for example, to reproduce themselves. The necessary principles were first stated by the mathematician John von Neumann. These automata could be sophisticated and versatile enough—like those found on the planet Regis III in Lem's novel *The Invincible*—to build cities, to exploit surface deposits, to render habitable the most hostile zones of the Solar System: our neighbouring planets, the asteroids, a Dyson shell. Dependent only on ores and solar energy, they could establish colonies to meet the needs of humanity on the Moon, on Mars, or in space.

Dyson gives us a vision of the Solar System in which biological technology and 'living' automata are in use, in the following quotation:

> Taking a long view into the future, I foresee a division of the solar system into two domains. The inner domain, where sunlight is abundant and water scarce, will be the domain of great machines and governmental enterprises. Here self-reproducing machines will be obedient slaves, and men will be organized in giant bureaucracies. Outside and beyond the sunlight zone will be the outer domain, where water is abundant and sunlight scarce. In the outer domain lie the comets where trees and men will live in smaller communities, isolated from each other by huge distances. Here men will find once again the wilderness that they have lost on Earth. Groups of people will be free to live as they please, independent of governmental authorities. Outside and away from the sun, they will be able to wander forever on the open frontier that this planet no longer possesses.[72]

Rebuilding galaxies

Dyson's concepts opened up new dimensions in scientific speculation. They were stimulated by *The World, the Flesh and the Devil*, the first book of the British physicist J. D. Bernal, which he wrote in 1929—a work which has influenced many science fiction writers, such as Olaf Stapledon and Arthur C. Clarke. Even then, Bernal was suggesting the building of space stations, totally independent economically, which would take in stellar energy from the stars on their travels or even tap stars directly. Bernal's vision is reflected in Clarke's *Rendezvous with Rama* (1973), and Dyson's shell, modified into a ring, found its literary echo in Larry Niven's *Ringworld* (1970).

Let us suppose that a civilization may have the capacity and the motivation to apply astrotechnology. How far can the astroengineers of a supercivilization go, and what would the Universe look like afterwards? To answer this question we need to be able to estimate how rapidly a highly developed civilization can spread. Any 'reasonable' nuclear-electric system of spaceship propulsion will bring it at least to 1 per cent of the speed of light: thus, a light-year will be covered in 100 years. With rapidly growing technology, astroengineers will have crossed the Galaxy, 100 000 light-years across, from one end to the other in 10 million years—a relatively short time, compared to the age of the Galaxy! Perhaps too long a time compared to the lifetime of a technological civilization. But let us assume this long lifetime. What does the Milky Way look like after this stellar journey?

What would a supercivilization do with a galaxy to make the greatest possible use of it in its urge for expansion? Planetary systems could be altered as described in the previous section. And perhaps the astroengineers will find that there is too little planetary material in the Galaxy. In that case they will take material from the stars themselves. The way to do this is to move stars around so that they get in each other's way. If a star is involved in a collision or a near-collision with other stars, one of the following will result: either the stars will throw enough material into space which will condense, cool, and be available to be processed, or the stars will collide and unite to form a larger star. But by collision with other stars a very massive star will eventually be formed. This goes through its nuclear burning stages relatively fast, at least considerably faster than a low-mass star, and will then lose mass slowly or explosively. While the Sun needs several thousand million years to convert its hydrogen to helium by thermonuclear processes, a giant star with 30 solar masses will take less than a million years. A galaxy 'tamed' by a Type III civilization, which has been using such processes for a long time, would distinguish itself from an untouched 'wild' galaxy by the following astrophysical properties:

- Intensified infrared radiation.
- More mass between the stars compared with the mass in the stars.
- Frequent stellar collisions between stars.
- An exceptional proportion of short-lived giant stars, together with a lack of ordinary dwarf stars (like the Sun).

In summary, we can say today that, so far as we can judge, astrotechnology is not impossible. Interstellar activities on this scale, do not, at any rate, contravene the laws of physics. Thus, astrotechnology is more a question of motivation and the lifetime of a civilization than a question of physics. Whether anyone has yet turned to the application of this astrotechnology is a matter we will investigate in the next section.

The search for artificial infrared stars

At the time of Dyson's first suggestion (1960) infrared astronomy was in its infancy. Today, 20 years later, there are already systematic catalogues of the infrared objects in the sky. Of course no precipitate conclusions should be drawn from the discovery of a strong stellar infrared source. Dyson himself warns: 'The discovery of a strong point source of infrared light would not in itself imply that an extraterrestrial intelligence had been detected.'[73]

Astronomers took Dyson's proposal seriously. Today many types of infrared observation are available.[73] Luminous supergiant stars with intense infrared radiation have been found in our immediate vicinity, and their infrared radiation actually has its maximum at the desired 10-micrometre wavelength. The interpretation has hitherto been that this radiation comes from silicate grains (sand) near the stars or the surrounding interstellar dust. Other strong infrared sources occur in connection with clouds of (ionized) hydrogen. It seems that in these cases the radiation from stars within the clouds, which also ionizes the hydrogen, is absorbed by the dust in the clouds and re-radiated in the infrared. But these stars are not visible optically themselves; they are completely hidden by the surrounding clouds and are therefore known to astronomers as 'cocoon stars'. In addition to the survey of starlike sources, the infrared radiation of whole galaxies is being studied. This gives at least indirect information on the presence or absence of Dyson civilizations, since galaxies which have been tamed by civilizations should above all display strong infrared radiation. Interestingly enough, galaxies have been found which radiate more in the infrared than in all other wavelengths together. Among these were the galaxy M82 and some of the so-called Seyfert galaxies. Naturally, our own backyard, the Milky Way, has also been checked for suspicious heat radiation. But the background radiation of the Milky Way, found at 10 micrometres, is probably emitted by the diffuse, uniformly distributed interstellar dust. The experts consider that the infrared sources hitherto observed are without exception stars in a very early stage of development: either the cloud is just contracting to form a star, or the young star has just been born and is blowing the remaining dust cloud away—as with the cocoon stars.[74]

Clearly, infrared radiation alone does not yet provide evidence of Dyson civilizations. Dyson therefore sketched out what, in his opinion, the future steps should be, in a lecture in 1972:

> Since I made the suggestion, infrared astronomy has grown by leaps and bounds, and hundreds of interesting infrared sources have been discovered. There is no reason to suspect any of the sources so far discovered of being artificial. Almost all of them have luminosities hundreds or thousands of times greater than the sun. Most of them are probably shells of dust surrounding very bright stars whose radiation keeps the dust warm. One possible model for an artificial source would be a long-lived star whose luminosity is necessarily not much greater than that of the sun, surrounded by a shell of habitable planetesimals. We shall only have a chance of detecting and identifying an artificial source of this kind, when the infrared scanning of the sky has been extended to objects hundreds of times fainter than the bright sources so far discovered.[75]

In the meantime, this has been done. Since Dyson's remarks were made, a good deal has happened in infrared astronomy. In particular, observations have been increasingly made from above the atmosphere (Figure 55). In the years 1971 to 1974, by means of high-altitude research rockets, 78 per cent of the sky was catalogued at a wavelength of 2.7 micrometres, and the positions of 2000 infrared stars were determined with an accuracy of 3 seconds of arc. The operation was carried out by the Geophysical Research Laboratories of the United States Air Force. Finally, in 1977, satellite observations produced the *Air Force Equatorial Infra-red Catalogue*, which covered positions between 10 degrees north and 10 degrees south of the celestial equator. The limiting sensitivity established a

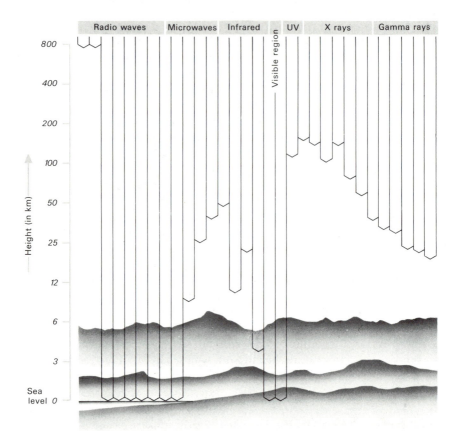

Figure 55 The Earth's atmosphere screens the surface of the planet from much of the electromagnetic spectrum. Cosmic radiation only reaches the Earth's surface through two 'windows', at radio wavelengths and in the visible region. The arrows in the diagram indicate at what heights above sea-level radiation from space is absorbed.

record: 'With these sensors it would be possible to register the heat emitted by a hundred-watt red lamp at a distance of 15 000 kilometres.'[76] Sources already known by their optical radiation were detected, but also new ones hitherto completely unknown: for 40 sources astronomers could find no image at the corresponding point in the sky on the sensitive photographic plates made with the Mount Palomar telescope. These are probably the faint sources known as Mira variables. Their light is dimmed by a gaseous envelope, and only the heat radiation penetrates to the outside. Even after these investigations, the most complete and sensitive yet made, it must be stated that no sign of Dyson civilizations has been found. But needless to say, much more astronomical work must still be done in this area.

Where is everybody?—the cosmic dilemma

Up to the present *no* astronomical observations have yielded evidence of extraterrestrial technology. Perhaps, therefore, all the considerations in the foregoing sections are pointless. Perhaps there are no supercivilizations. Perhaps civilizations inevitably destroy themselves or sink into decline when they pursue the path of technology which began with the industrialization of the nineteenth century. Is it only by evolving without a technology that a society has a chance to survive over cosmological periods of time? But then we would never detect their presence: why should they build radio telescopes or dismantle their planets?

We may formulate the fundamental dilemma of the present search for extraterrestrial *technological* intelligence as follows. On the one hand, many astrophysicists and biologists believe, on the basis of their probability calculations and laboratory experiments on the origin of life, that the Milky Way (like other galaxies) is populated by technological civilizations. A typical estimate is 10 million! On the other hand, no indications have yet been observed of the activities of intelligent creatures in the Cosmos other than on Earth.

This dilemma raises a difficult question, and one which is relatively independent of what we do not yet know about supercivilizations, and what astronomical phenomena we, as mankind, actually hope to discover with our telescopes. In the words of Sebastian von Hoerner: 'In just the same way, some untouched aborigines could not guess what cities look like; but if one day they came upon a city they would immediately recognize it as being something artificial, produced by higher beings, and not a natural growth.'[70]

There are several answers to the question: 'Where is everybody?' (One of them, the enigmatic UFOs, will be discussed in Part III.) Many scientists are concerned with the apparent paradox presented by this fundamental dilemma, especially the astronomers and theoreticians Freeman J. Dyson, Sebastian von Hoerner, Michael H. Hart, and Michael D. Papagiannis.

As long as non-classical astronomy remains in its formative stages—radio, X-ray and gamma-ray astronomy only began in the last 20 years—we can fairly say that perhaps we have still observed far too little. Despite the great advances in these areas our information may still be insufficient to refute the existence of supercivilizations by the fact that we still lack evidence for them. So long as this holds good, the remark of Martin Rees applies: 'Absence of evidence is not evidence of absence.' So long as the number of astronomical objects which have been studied does not comprise a statistically representative sample, all conclusions drawn from it concerning the absence of extraterrestrial technology are unsupported by the facts, and the opposite conclusion—the existence of supercivilizations—may be just as valid. However, there remain only two possibilities: either we observe no astrotechnology because we have not yet looked for it thoroughly, or we do not see it because it is not there. In the first case we shall have to wait patiently and continue to pursue our astronomy, which will happen in any case.

But it may be possible (the second case) that there is no other technology in the Galaxy discoverable by radioastronomy. 'I have the feeling that if an expanding technology had ever really got loose in our galaxy, the effects of it would be glaringly obvious. Starlight instead of wastefully shining all over the galaxy would be carefully damned and regulated,' Dyson reluctantly admits.[69] And precisely when energy is being economically used, it will be returned to space only in the form of useless heat radiation. This should make a civilization observable above all as an infrared source.

And at least one civilization must have done this if so many technological civilizations have actually appeared in the Galaxy. Are we the only technological civilization in the Galaxy? This question may cause the terrestrial scientist of the twentieth century to smile. He has learned from the history of science since Copernicus that the Earth, and probably also its inhabitants, represents nothing special in the Universe. Of course it cannot be ruled out that mankind is an exception, one of the first of its kind, and that no other intelligent creatures have yet had the capacity or the time to undertake something which we can detect here on Earth. Let us therefore consider more carefully some of the various possible consequences of the 'cosmic dilemma'.

Three hypotheses

Hypothesis One: *There are no extraterrestrial, highly developed technological civilizations.*

This hypothesis permits the existence of any number of extraterrestrial and also intelligent life-forms in the Universe with civilizations of Kardashev Type lower than I. However, it strongly contradicts the prevailing estimates of the number of intelligent civilizations in the Galaxy with an advanced technology (N). The estimates of N—up to hundreds of millions in our Galaxy alone—are largely derived from the assumption that life tends to expand into the whole of the space available to it, and that the technology required for this also makes interstellar travel possible.

Here I assume that space travel is possible (see Part III). But the assumption that life will fill all the available living space is open to serious doubt. This behaviour, which Papagiannis describes as 'observed', is certainly not an observation in the scientific sense.[77] At present it has only been experimentally tested locally (for small areas of the Earth) and for the early stages of organisms or groups. It is certainly established that life at first expands spatially, for variety initially requires space. But later, in the mature phase, for instance, we can conceive not only of an expansion in space, but also of an expansion 'inwards': if a society is interested in longevity, this may impose itself as the only route which can be followed. For mankind this 'inward' route might look something like this: in the phase of industrial growth and learning—which we shall soon have left behind us—mankind mainly accumulates material goods. When the planet Earth has been stripped bare for these purposes by the consumption of its raw materials, the post-industrial phase of society sets in. Here knowledge and information, rather than objects, increasingly become the most important commodities. A society which is primarily increasing its store of knowledge does not necessarily need further expansion in space to do it. Here a fundamental question arises: what will then happen to our allegedly unstoppable technological progress?

The stimulus for further technological evolution may be lost simply by a change in the principal interests of a society. For example, we cannot even predict how long our present desires, our technological aims, will continue to be pursued. A society which is able to survive for a long time must have developed very effective methods of protecting itself from all suicidal crises and accidents. In order to achieve this, a society in the process of stabilizing itself must have subordinated itself to behaviour governed by reason. In a society capable of surviving, one which is stabilized, and therefore has a constant population, there is no longer any great stimulus to space travel, the expression of an urge for expansion and emigration.

Of course, such a stabilization and inward expansion by no means rules out contact with other civilizations. Radio communication is easier than travel, and will also bring a stabilized civilization the stimuli which will protect it from decay. 'Interstellar expansion would then be spiritual, not corporeal.'[70]

Finally there is the possibility that a society is unable to stabilize itself. Then, in the opinion of Sebastian von Hoerner, it is likely to go through three critical phases: population explosion, self-destruction and finally genetic degeneration. Today we have a reasonable understanding of the first two types of crisis: they now present the most serious threats to the survival of mankind. The third crisis is probably some centuries away. The likelihood of *not* overcoming these crises is high, and accordingly—if analogous crises also occur in other civilizations—technological evolution will very soon be halted or reversed.

But in general there is a problem with all these attempts at explanation, a problem which none of them can escape.[78] To account for the absence of technological civilizations, the explanation must apply to *every* extraterrestrial race, irrespective of its biological, psychological or political structure; and to *all* the phases of the history of each civilization, at least to the period since it reached the age of space travel. The following hypothesis avoids this criticism.

Hypothesis Two: *Our theories about the origin of the evolution of life are wrong; we may be alone.*

This thesis is in fact harder to swallow than the first. Experiments in prebiotic chemistry suggest that life is a common phenomenon in the Universe.[79, 80] We have just learned to complete the 'Copernican revolution in biology' and to assume that the greatest variety of intelligent life-forms may occur in the vicinity of many of the 100 000 million (10^{11}) stars of the Milky Way. Nonetheless, it could well be the case that there are only a few technically advanced societies, or only ourselves. In order to find the error, the 'scapegoat', we must turn back to the formula proposed by Drake and Sagan, by means of which we calculated N, the number of societies which have developed technologically to the point of being capable of interstellar communication (see Part I: 'The number of advanced cultures in the Galaxy', page 87).

If our second hypothesis is to be true, the number N (up to 10^8 in the Galaxy) must be reduced by the enormous factor 10^6 or 10^7. If this hypothesis *is* true, an error of at least six orders of magnitude has been made in one or more of the factors. Let us examine which of the eight factors may be involved. The data which are relatively the most trustworthy are those derived from astronomy alone: R, f_g, f_p, and n_e. Astrophysicists think that these factors, taken together, may be 'out' by a factor of at most 10 to 100. The factor f_a can also be given fairly reliable limits: not all intelligent societies will at once develop a higher technology. But 1 in 10 or at least 1 in 100 can be expected to take this further step once it has become an intelligent civilization. The weak point we seek seems to lie in the two remaining factors. These factors, indeed, are easier to guess at than to calculate: f_l, the relative number of planets orbiting within the habitable region of a star on which life actually appears, and f_i, the relative number of planets on which biological evolution finally leads to intelligent life. These factors conceal the complex interplay of the theory of young planetary atmospheres on one side, and of the intricate biochemistry of the origin of life on the other. These processes are still largely unknown, and only quite recently have halfway realistic computer simulations been carried out for some of their aspects.

These calculations, for example those of the NASA scientist Michael H. Hart, indicate that the life-supporting zones around the stars are much narrower than had previously been thought (see Part I: 'The life-supporting zones of the stars', page 32). Given a minute reduction in the Earth–Sun distance, the atmosphere of the planet will heat up by the greenhouse effect, as on Venus. At a greater distance the oceans will freeze after all the clouds have fallen as rain and produce a glacier climate, hostile to life, as on Mars. Life therefore appears much more rarely. This could reduce the factor f_l to a thousandth (10^{-3}) of the value previously estimated.

The other questionable factor, f_i, for the development of intelligence in a life-form, is usually set equal to one. The laws of evolution—it is generally assumed—

should always develop intelligence in a race which is optimally suited to its environment and has consequently prevailed over other varieties of life. But one of the critical phases in the scheme of biological evolution is the transition from primitive unicellular organisms to more complex many-celled forms. On Earth, this stage of development lasted longer than all other phases of evolution: 3000 million years. Therefore it may well represent a critical barrier, at which biological evolution most frequently 'sticks'.

During these 3000 million years the Earth's atmosphere changed radically from an envelope rich in carbon dioxide to an atmosphere containing large amounts of oxygen. For this to happen, several factors had to harmonize exactly: water could not freeze, and the ozone layer had to keep most of the ultraviolet radiation from the Earth's surface—but only after the ultraviolet rays had assisted with the production of simple biological molecules. Besides this, the size of the planet, the inclination of its axis and the eccentricity of its orbit (its deviation from a circle) probably have all to be just right, to guarantee exactly the proper variations in temperature over thousands of millions of years; though there have not yet been any scientific studies of the effects of changes in these properties on the transition to an oxygen-rich atmosphere. However, in the opinion of Papagiannis an additional factor should now be included in the mathematical formula for N, which would take into account the critical physical and chemical characteristics of life-bearing planets. If conditions lead to an oxygen-rich atmosphere on only one planet in a million, this could explain why so few (or no other) technically advanced civilizations have appeared in the Galaxy; and this would probably make estimates of the number of planetary systems with technically advanced societies too optimistic by several orders of magnitude. But if the number of candidates to be considered is drastically reduced, the astronomical observations which have hitherto been made are of correspondingly greater importance. I therefore consider that the current negative results of the search for artificial sources constitute weightier evidence than other astrophysicists suppose.

As an alternative, many scientists studying the problem of extraterrestrial intelligence believe that 'they' are trying to communicate with us, and we only need, in effect, to pick up their signals. Since this idea is in violent contradiction with all observations, some authors of science fiction stories, and also some serious scientists (for example, the Harvard physicist John A. Ball and recently E. M. Jones) have thought up a sophisticated answer, which allows the existence of any number of supercivilizations in the Galaxy, but in such a way that we can observe nothing of them.[81,82] We are artificially shut off from our galactic environment, enclosed in a sort of wildlife reserve or zoo. This is the content of:

Hypothesis Three: *Mankind is developing on Earth in artificial isolation from the Galaxy, screened off as though in a zoo which supercivilizations have constructed so that we cannot perceive them.*

Are we the occupants of a zoo?

The seven thousand and fifty-fourth session of the Galactic Congress sat in solemn conclave in the vast semicircular hall on Eon, second planet of Arcturus . . . The president delegate's voice boomed out then: 'Delegates! The system of Sol has discovered the secret of interstellar travel and by that act becomes eligible for entrance into the Galactic Federation.'

A storm of approving shouts arose from those present, and the Arcturan raised a hand for silence.

'I have here', he continued, 'the official report from Alpha Centauri, on whose fifth planet the Humanoids from Sol have landed. The report is entirely satisfactory and so the ban upon travel into and communicating with the Solarian system is lifted . . .'[83]

We have not yet come as far as the scene described in Isaac Asimov's early short story *Homo Sol*. But this is the way events in the Galaxy could turn out if the zoo hypothesis is correct.

But we should first make clear what level of scientific validity underlies these three hypotheses, which, to begin with, are only speculative possibilities. At least in principle, a scientific assertion must be open to disproof. What the scientist is looking for is a statement he can test for its correctness practically in his own laboratory. Claims regarding astronomical objects far away in space are more difficult to check: our contact with these objects is only indirect, through their radiation, which, for example, we can photograph, and only then begin to interpret by means of models and theories. But where numerous differing interpretations agree we can still arrive at an acceptable, scientifically tenable statement.

The difficulty is far greater with theories constructed in just such a way as to be, as far as possible, irrefutable.

So it is with the zoo hypothesis. It is precisely so constructed that we cannot perceive the existence of extraterrestrials, at least from the Earth. There are only two ways in which this hypothesis can be refuted. Either we do indeed succeed, using Earth-based astronomical techniques, in tracking down extraterrestrial societies, or we set out ourselves on the long journey through the Galaxy and thus break down any existing barriers between Earth and Galaxy. Then we either find the 'watchers' or establish by a thorough search that we are the only technological culture. Hypothesis One is on rather better ground. As astronomical searches become increasingly thorough, we shall already reach in the 1980s a degree of detail which will make it less and less likely that we have missed any technological civilizations within the Galaxy. Non-technological but still intelligent life-forms, which Hypothesis Two would exclude, can in fact only be refuted if we set off and scour all potential life zones—in principle the same effort as for testing the zoo hypothesis.

Only civilizations at the highest level of technological development are of significance in the zoo hypothesis. They will, in a sense, 'supervise' the Milky Way (or larger parts of the Universe). If we may draw an analogy with the behaviour of peoples on the Earth, the more technologically developed groups will in course of time either assimilate backward civilizations, destroy them, or bring them under their control in some other way. Technical advance can be defined as increasing capacity to control one's environment. At our own level of technology we already influence almost all life-forms on Earth, from elephants to viruses. But we do not always exert the influence available to us, our power to control, in a direct way. For a wide variety of reasons, for example, we set up nature reserves or zoos, in which certain life-forms—relatively uninfluenced by man—are practically left to themselves and their environment, to develop undisturbed. For the owner of a zoo, the ideal zoo would be an enclave in which the life-form (or forms) had no contact at all with the 'outside world', and knew nothing of the existence of the proprietor.

According to the zoo hypothesis, the obvious absence of any contact between 'them' and us can only be understood if 'they' are deliberately avoiding us and have arranged our surroundings like an ideal zoo.

It is possible that we really are in a laboratory situation of this kind, like amoebae under the microscopes of the lords of the Galaxy. 'There is a serious question about whether such societies are concerned with communicating with us, any more than we are concerned with communicating with our protozoan or bacterial forebears,' argues Carl Sagan.[84]

The zoo hypothesis is not very flattering to human vanity—indeed, the picture is oppressive to many people. Questioned on this possibility, Martin Rees replies (personal communication, 1977) without hesitation: 'Not a pleasant picture—I should not feel happy in such a Universe.' Psychologically it would certainly be most pleasant for us if we could think that even more advanced supercivilizations wished to make contact with us, or would at least attempt it as soon as they knew of our existence on Earth. However, in the history of science psychologically unpleasant hypotheses have often enough been shown to be scientifically correct.

But yet abundant reasons can be imagined to explain *why* a supercivilization would want to keep mankind in a zoo. Here we must ask ourselves what requirements might indeed make a civilizing relationship desirable, but not until later, in the sense that contact would be meaningless, at present, since we could not yet supply what was desired. As already argued several times, a supercivilization would hardly be interested in our technology (except to establish a basis of understanding), but at the most in our cultural achievements and historical developments on Earth. This information is a commodity which is constantly on the increase. T. B. H. Kuiper and M. Morris consider that this civilizing process of adding to knowledge should not be arbitrarily disturbed by provisional contacts. 'Before a certain threshold is reached, complete contact with a superior civilization (in which their store of knowledge is made available to us) would abort further development through a "culture shock" effect.'[28]

Before we reached this threshold of maturity, we would not enrich the galactic store of knowledge, but only absorb it in a one-sided process. For example, much culturally unproductive time would be used in absorbing knowledge rather than developing it for ourselves.

If there are other goods on the Earth today which are of interest to extraterrestrials, they could easily come by them without any open contact. Rare elements or chemicals, even genetic material or individual animals or people, could be removed from the Earth without making much stir. Thousands of people 'vanish' every day. In this half-hidden manner an extraterrestrial society could maintain contact with the Earth, without the possible risk of cultural shock—in order to investigate human reactions experimentally, or to direct evolution in a given direction, or even to save the Earth from a great self-destruction.

These thoughts already run close to those of the 'ufologists' and the speculations promulgated by R. Charroux and Erich von Däniken, which will be more fully discussed in the last section of the book (page 273). But first let us examine a special aspect of the zoo hypothesis: whether extraterrestrials may not already have entered our Solar System.

Extraterrestrials in the asteroid belt

According to our present observations, there is no evidence of any kind of organism in the Solar System outside the Earth. This is the result of all the voyages of exploration made by space probes to Mars and Venus. Nor have the American satellites Voyager 1 and Voyager 2 (launched in August and September 1977—see Part III: 'The Galactic Club', page 235) on their flights past Jupiter and Saturn yet provided any observations to contradict this, and are unlikely to do so as they speed on to Uranus. Before the Mars landing this planet had been considered to offer the best chances for organic life. Of course there is still the possibility, following the zoo hypothesis, that the extraterrestrials, should they have penetrated the Solar System themselves or with probes, are 'keeping a low profile', or that we have not yet confirmed their presence.

Either way, it does no harm to ask: if our Solar System has actually been visited and settled and is now watched over by extraterrestrials, where can they have established themselves? Earlier, our neighbouring planets were suspected as harbouring other types of life (a large part of the literature of classical science fiction drew on this), but the Moon, Mercury, Venus and Mars are now clearly eliminated for these purposes—and these were the best candidates! Besides, we could scarcely expect to find extra-solar creatures on these planets. Unlikely to be suited to life on the planets of the Solar System, furnished with an artificial environment created by themselves (spaceships) for a sojourn in space, it would be more probable that the extraterrestrials would continue to live in their spaceships once they reached the Solar System, merely 'parked' in suitable orbits about the Sun. 'Suitable' here means at such distances from the Sun that the best supply of energy through sunlight and raw materials is assured.

The asteroid belt, also a future area for terrestrial settlement, seems to be best suited to these demands. This is an old idea, which is still frequently taken up by scientists.[77] Important materials could be obtained directly from asteroids metres to kilometres in size. The danger of the destruction of space stations by meteoroids is evidently smaller, too, than has previously been thought. The planetary probes Pioneer 10 and 11 did not encounter any more meteoroids in their flight through the asteroid belt than anywhere else in the Solar System. Both passed through the ring of dust grains and rocky fragments unscathed.[71]

In the speculative technical literature it has occasionally been suggested that the asteroid belt may be the end-product of the violent destruction of a planet which, according to the Titius–Bode Law for the distances of the planets from the Sun, may have orbited the Sun at this distance. Were astroengineers at work here, perhaps with the object of exploiting its metals? Was this the scene of a battle between the stars in the dim past? Hardly. The planets were formed from a gas cloud—the primitive Solar Nebula—which had flattened itself into a disc by its rotation. The disc first separated into rings, in which dust grains, and, through successive collisions, progressively larger bodies, the so-called planetoids , were formed. This mechanism produced the inner planets, Mercury, Venus, Earth and Mars. It was different with the outer planets. In Jupiter and Saturn, which consist principally of hydrogen and helium, gas from the primitive Solar Nebula must have continued to collect about planetoid nuclei until it collapsed on to planetary nuclei. The asteroid belt, lodged between the small inner planets and the massive outer ones, between Mars and Jupiter, has never quite succeeded, according to this theory, in becoming a proper planet. It 'stuck' at the stage of small, colliding planetoids. The heavy planet Jupiter was (and still is) responsible for this: the disturbing influence of its gravitational field impeded the formation of a proper planet (see Part I: 'The birth of the planets', page 19). I. S. Shklovskii's maxim is also valid here, that 'every event must be regarded as natural until the contrary is proven'.[18] But certainly a spaceship a few kilometres in size would be very difficult to distinguish from the Earth among hundreds of thousands of 'natural' asteroids of similar size. Beamed radio contacts between inhabitants of the asteroid belt and stations outside the Solar System would scarcely be detectable on Earth, at least not without considerable effort. The scattered radiation, direct from their telescopes or reflected to Earth from the outer planets, would be extremely weak.

If we start from the assumption that there is at least one supercivilization in the Galaxy and that it operates space travel (this assumption is part of the zoo hypothesis), then, to summarize, there are at least three reasons for paying more attention to the asteroid belt:

• In colonization of the Galaxy by a supercivilization it is very unlikely that a system as 'attractive' as the Solar System would be overlooked: the Sun is a relatively young, long-lived and 'well-behaved' star of the Main Sequence, with many planets, moons, asteroids and meteoroids, surrounded by a whole cloud of many thousands of millions of comets.

• Civilizations with space travel will probably make their journeys in space colonies, which draw their energy from stars and their raw material from planetoids of low surface gravity.

• The asteroid belt is an inconspicuous place in which to capture sunlight for their own needs, to exploit raw materials and at the same time to establish themselves in a space colony.

The asteroid belt must certainly be given much closer attention—and not only because of the remote chance of coming on extraterrestrials there. However, a search for radio waves, infrared analysis and finally a complete photographic examination of the asteroid belt with space probes would at least settle this question once for all.

An intruder in the Solar System?

At the beginning of November 1977 an event occurred which seemed at first glance to support the zoo hypothesis. The American astronomer Charles Kowal discovered a new celestial object between the orbits of the planets Saturn and Uranus with the reflecting telescope on Mount Palomar in California.[85] An unknown body, several hundred kilometres in size, on course for the Earth? Or were extraterrestrial observers sitting there keeping watch? The guesses about the origin and nature of the object which had so unexpectedly appeared were not confined to astronomers.

Certainly this was not the discovery of the long-sought 'tenth planet' (as the first announcements had claimed), which, according to the law for the planetary distances from the Sun, should be still to be found outside Pluto, the outermost planet. Because searches in this direction have been failures it is now thought that there is no tenth planet in the Solar System. Also, the new object is much too small, too small indeed for its size to be directly measured. At about 300 to 400 kilometres it has about a tenth of the Moon's diameter, if we assume that it reflects sunlight as the rocky surface of the Moon does. If the surface is darker it could be as large as 600 kilometres. The planetary expert Brian Marsden of Harvard University commented: 'It's the smallest body ever discovered so far away from the Sun.'[85]

In about 25 years (with an orbital period of 49 years) its highly elliptical orbit carries Object Kowal from aphelion near Uranus to inside the orbit of Saturn.

Saturn and Uranus, the large neighbours of Kowal's object, take respectively 29 and 84 years to go around the Sun. By calculating backwards over the orbit it was found that Object Kowal must have pursued its very stable path for at least

2500 years. The calculations were quickly verified: the body could be identified with their help on old photographic plates taken in 1969.

At present it is not completely clear to what category of heavenly bodies Kowal's object should be assigned, or how it comes to be where no such object had been expected. As far as its size is concerned, it could pass for an interstellar spaceship. If the orbit of this body should suddenly change in the next few years, this would indeed be a strong indication that it is artificial. But so long as Object Kowal pursues its predicted orbit like a planet, the spaceship hypothesis cannot be judged with absolute certainty by astronomical methods. It is unlikely, however, that we are dealing with an asteroid. Asteroids move far nearer to the Sun, chiefly between the orbits of Mars and Jupiter.

In fact the only remaining possibility is to regard Object Kowal as a comet. But this explanation, too, does not fit too well: Object Kowal is moving almost in the orbital plane of all the planets, with an inclination of 3 degrees to the Earth's orbit; this is not typical of comets, which approach the Sun from the greatest variety of directions. The orbit suggests that Object Kowal originated in the neighbourhood of the outermost planets, Uranus and Pluto. According to the theories of the origin of the Solar System, the outermost planets were formed from small bodies made of frozen gases, ice and dust—so-called cometoids—in the same way in which the inner planets, including the Earth, were formed from planetoids. Object Kowal, then, is probably a relic of the history of the formation of the outer planets.

Why was Object Kowal not discovered earlier? For one thing it moves very slowly and reflects very little sunlight. The Sun would have to be moved to a distance of 15 000 light-years, halfway to the centre of the Galaxy, to appear as bright to us as Object Kowal. (The actual Earth–Sun distance is 8 light-minutes, which represents 150 million kilometres.) Kowal, who had already discovered Jupiter's thirteenth moon in 1976 (and since then possibly a fourteenth), actually came on his Object Kowal rather by accident, while he was searching this empty corner of the Solar System for distant, faint comets. Kowal believes that it is not out of the question that there are yet more mini-bodies of this kind in distant and still less thoroughly searched regions of the Solar System.

Consequences of contract: the influence of interstellar communication on human progress

In a sense, some of the 'consequences' of interstellar relations—and currently the only ones—appear even before the first contact is made. They arise from the profound consideration given to all the things which could possibly happen once it *has* actually happened. Or from reactions to experiences that our contemporaries have had with unidentified flying objects—the expression of an attitude of expectancy fed by uncertainty. Unexplained phenomena, threats from outside a society, have always led to demonstrations of mass psychology in the history of terrestrial culture. From sociological research we know that societies often construct those hostile figures and corresponding taboos needed for the political stabilization of their particular form of society. The stronger the psychological pressure on individuals within the group of these taboos and norms of thought and behaviour, the greater becomes the need to project these tensions outwards on to minorities within their own society or directly on to other peoples. The occasional spectacular reactions to alleged contacts with extraterrestrials should above all be seen in this light. It is only away from these reactions to possible or actual contact, coloured by mass psychology, that the possibilities of the various consequences of contact can be given rational limits, albeit using anthropomorphic judgement, influenced by the unconscious shared anxieties of mankind. In the same way, the hypotheses, especially those propounded by science fiction authors, range from destruction by war to the attainment of immortality.

'As soon as a civilization has established contact with another world its life expectancy would be greatly increased, since the knowledge that others had been able to survive the crisis, and perhaps some guidance as to how this was attained, would make the new member of the galactic community better able to solve its problems . . . ,' Sebastian von Hoerner optimistically concludes.[70]

Before we get that far, some problems may still present themselves. Perhaps 'extraterrestrial intelligence' is easy to identify, but we are distinctly limited in our choice of intelligent forms of life. Even if contact has been established, actual communication may well break down because we do not know what the other race regards as 'intelligence'; and the extraterrestrials may have the same difficulty. Consequently, an interstellar radio signal will not be as easy to read as a telegram sent from house to house. Philip Morrison also thinks this: 'The recognition of the signal is the great event, but the interpretation of the signal will be a social task comparable to that of a very large discipline, or branch of learning.'[86]

Accordingly mankind, at first compelled to put hard work into the reconstruction and translation of the message, would be shielded from the dreaded and powerful cultural shock. But in the long run, contact with extraterrestrial civilizations could have a profound effect on the future development of mankind.

If successful contact is actually achieved, we shall scarcely be the first civilization to have done it, even on an optimistic view. The current idea is that the societies of the Galaxy have been linked in a permanent exchange of information for thousands of millions of years. As a consequence, more and more information would be collected in the course of time and put at the disposal of the newcomers through the galactic information service—a growing galactic cultural heritage. It would comprise the history of unnumbered planets and races which have already perished, and astronomical data from the early days of the Galaxy and its neighbouring galaxies. Old and young civilizations would exchange their respective learning peacefully and without excitement, and would grow, flourish and decline in the process (see Part III: 'The Galactic Club', page 235).

There are many who believe that this almost divine knowledge would cure the world of its ills. We would learn how to master our current crisis, protect mankind from self-destruction, stabilize our society and prevent genetic degeneration. Furthermore, cultural and political attitudes would be put in perspective and modified. This is the fable of a galactic patriarchate. But in the light of the astronomical facts and theories of biological evolution which will be further summarized in the final section of this Part (page 222), this seems to me to be the childhood dream of a not yet mature human society which yearns for refuge from its fears of a self-destructive future.

Monsters conquer the Earth

Positive consequences of the encounter between different civilizations are conceivable, and may actually come about—to some extent at least. But these considerations derive from an assumption—that of mutual goodwill and the existence of common interests between alien societies. Since we know practically nothing beforehand, we should make do with as few assumptions as possible, and above all avoid constructing a happy world of 'galactic societies' as a projection of Earthly hopes of redemption.

Fears of an opposite nature are even more widespread: since 'they' are probably superior to us, they have nothing else in mind but to conquer us, to strip us of everything and to destroy us. When all is said and done, this would not go against all our experience of the relations between human societies. The history of human civilizations consists principally of records of conflict—war, plundering and subjection—between nations, groups and races. At the same time, this would impute to cosmic civilizations a type of behaviour which is probably sought-after so emphatically and systematically nowhere but on Earth. The popular idea that contact with extraterrestrial civilizations will inevitably have negative consequences derives essentially from mankind's own aggressive behaviour in the past. Usually, it starts from the supposition that extraterrestrials will visit us in person by spaceship. Besides total destruction, the scenario of negative consequences is essentially: invasion, subjection and exploitation, secret infiltration and cultural shock.

The pictorial plaque on board the space probe Pioneer 10 was criticized, among other things, because it 'betrayed' the position of the Sun (Earth), and would thus expose us to the colonizing desires of other civilizations. Perhaps it was just such fears, nourished by the brutal story of colonization on Earth, which even brought legal experts into the picture. We may be surprised or amused, but in 1970 there appeared a book, *Relations with Alien Intelligences—The Scientific Basis of Meta-Law*, by the Austrian 'space lawyer' Ernst Fasan.[87] In this he lays down legal principles by which cosmic civilizations are requested to deal with one another. It reads like a Magna Carta for the members of a galactic League of Nations: 'No partner of Meta-Law [the law which is applicable between interstellar civilizations] may demand an impossibility; all intelligent races of the Universe have in principle equal rights and values; every partner of Meta-Law has the right of self-determination; any act which causes harm to another race must be avoided . . .' Fasan really believes that his legal recommendations are in no way anthropocentric, but are by necessity inborn in all intelligent races.

But what if emissaries from another civilization, with a cultural past perhaps 100 000 years or so older, actually were to approach the Solar System as colonists or conquerors? I leave it to the reader to imagine whether we could obstruct any plans the 'invaders' might have by assuring them that we had already prepared the basis of a legal and democratic partnership. What would the Spanish conquistadores have said if the Aztecs had met them with a similar proposition?

But how should we actually assess such dangers? Space travel is certainly possible, but it will still be an expensive business, even for a highly developed society, if undertaken on a large scale (see Part III). Even the inhabitants of these will only set off on long journeys in the event of an extreme crisis. The study for the American Project Cyclops concludes: 'It is our opinion that we can neglect a search for additional living space as a motive, since any race that is capable of interstellar emigration will long since have solved its population problems by other means.'[35] On the other hand, a threat to the living space of a society, such as an impending nearby supernova explosion, would be good grounds for mass emigration. Whether the forced emigrants would then positively seek out alternative quarters on an already over-populated planet is very doubtful. Instead of effecting a violent entry to inhabited planets, it will be more economical for them to seek out uninhabited regions of space compatible with their form of life.

The other question is: would we remain 'hidden' if we discontinued all radio transmissions? In the long run mankind will not be safe from the possibility of discovery simply by 'keeping quiet' (unless it dies out in time). Even with a relatively small rate of expansion of only a few per cent of the velocity of light, and despite radio silence or space probes like Pioneer 10 without explicit indications from the sender, the Earth would hardly remain undiscovered. This would be particularly true if, on account of physical laws still unknown to us and the corresponding methods of propulsion, space travel were to be easier and cheaper than is now assumed.

Aside from an invasion of the Earth, aside from enslavement and life-destroying exploitation, a more sophisticated threat from an extraterrestrial culture is conceivable: if, after making friendly contact with us, they persuade us to undertake actions which later make it possible for them to gain control over mankind. This idea has been developed in a science fiction story. The 'aliens' offer us a medicine which cures all mankind's serious diseases. The cure, tested at first only on the dying, is successful; then everyone is inoculated. But when after a time some people happen to notice that their IQ is declining, it is already too late for mankind.

Threats of this kind—fictional as they may be on the whole—might be hard to ward off except by caution and commonsense; just as it is difficult to counteract the effects of a cultural shock which a contact with even the best-intentioned alien civilization might produce. The influence of a considerably more advanced race may be so powerful that the cultural development of the human race itself might simply be stopped. Relationships between societies on the Earth have almost always led to the domination of the weaker by the stronger, though this has usually been accompanied by personal contact and territorial annexation. Unfortunately we cannot orient ourselves to any example in history in which two cultures made contact with one another only through the medium of radio. On the level of interstellar contact, we clearly run no danger at all by receiving messages of any kind. Only when we rouse ourselves to send an answering signal is there a potential threat, though not an immediate one, given the long interval between transmission and reception. The mere reception of a radio message will scarcely spread panic and terror, as is occasionally suggested. Panic is an exceptional reaction, which occurs above all when altered conditions maintain their effects for some time. On the other hand, the discovery of an extraterrestrial civilization is a once-only event. To begin with it will certainly excite great popular interest, with occasional hysterical outbursts like those produced by the announcement of UFOs, only to die away when the public receives more detailed information. In the worst case we can always turn off the receiving telescope.

More important than the immediate reactions are the long-term changes produced by reception of a radio message. Mankind as a whole would be confronted in a unique way with another civilization. The very existence of an ancient race, not too different from our own, would mean that this civilization must somehow have survived its self-destructive crises (if it has had them) without renouncing all subsequent technical evolution. It is not entirely improbable that we might finally learn how mankind too can survive the crises of civilization of the twentieth and twenty-first centuries. So long as we receive no artificial signal, no comprehensible message, we shall be best employed in asking ourselves whether we desire contact with extraterrestrials in the first place, what information we expect to receive in such a message, what we ourselves would like to send, and what we actually promise ourselves from all this.

The astronomical search for artificial electromagnetic waves and traces of extraterrestrial civilizations is continuing, and in the coming years will, in one way or another, if not settle the issue, at least narrow it considerably. Then we shall either have come upon the existence of another civilization, or can rule out, with far higher probability than before, the existence of technological civilizations, at least in the Milky Way. 'To be part of this far reaching expression of mankind's desire to know is to partake of a dream that binds us temporally with our past and future, spatially with the cosmos, and culturally with our destiny.'[33] Then we shall know more surely whether the previous 'absence of evidence' was not 'evidence of absence', or whether for the foreseeable future mankind must remain content with its own company.

There have been, however, attempts to infer from the observed absence of extraterrestrial beings their general non-existence—contrary to Martin Rees's cautioning.[78] If extraterrestrial beings existed, runs the argument, their space-ships would be present in our Solar System—or, more strongly: since there are none here there are none anywhere.

Naturally, any attempt to construct a 'proof' of this claim will have holes in its argument as large and as numerous as those in a Swiss cheese. Nevertheless, it is quite a fascinating speculation, and competent reasoning along these lines should be given proper attention.[88]

Conclusion: we are (still) alone

The close of this chapter on interstellar communication will also bring to a close our detailed discussion of facts and fantasies on the subject. It is time to take both the theories of the origin of life and the available biological, palaeontological and astronomical observations for what they are; many speculations and fanciful proposals can then be discarded as adventures of the imagination.

Certainly, the 'Copernican revolution in biology' was justified. Life probably occurs in the Universe far more often than was thought possible some decades ago. There are many pieces of evidence suggesting this, as we have seen in Part I. On the Earth there are microbes living on the hydrogen sulphide of the deep-sea volcanoes near the Galápagos Islands; methane bacteria which consume carbon dioxide and hydrogen and produce methane, the third and possibly the oldest type of life on Earth, found in geysers and in the oceans; and finally organic chemical predecessors of life, far from stars and planets at tempera-tures of around $-200\,^{\circ}C$—the organic molecules in the gas clouds of interstellar space. The chemistry of these gas clouds produces, on the surface of microscopic dust grains, not only molecules familiar to us, but also exotic types which could not exist on Earth and were therefore previously unknown. About 50 molecules have hitherto been found, including amino acids and ethyl alcohol. And it is suspected that still more complex compounds remain undiscovered in space simply because our radio telescopes are not sensitive enough. In meteorites, however, we have found molecular fragments of the so-called nucleotides, whose relatives on Earth form building-blocks for the nucleic acids RNA and DNA. Although they do not yet represent an independent form of life, these end-products of chemical evolution may have provided the initial impetus for biological evolution on the young, recently cooled Earth. Fred Hoyle and Chandra Wickramasinghe hypothesize (see page 43) that even biological seeds of life, viruses and bacteria, have bombarded the Earth in cometary fragments and meteorites, and even today reach us undamaged by frictional heating as they plunge into the atmosphere.

It cannot be completely excluded, although it is unlikely, that a life-germ from the Universe became the initiator of the evolution of terrestrial life. (This idea, moreover, only removes the problem of the origin of life into interstellar space.) And it is indeed equally conceivable that the biological invasion of the Earth is still going on today. But the supposition that these hypothetical germs can have been responsible for many of the epidemics of history is a most audacious and implausible speculation for these two scientists to have put forward.[89] Within the Earth's biosphere all living things have the same genetic code, entrenched through a historical accident, the scheme for translating the information stored in the nucleic acids for the purpose of protein synthesis. But there is a vanishingly small probability that any non-terrestrial evolution would by chance develop the same code. Life-forms with different codes, however, could not influence one another biologically; at best they would be able to develop separately among themselves. But up to now no such life-forms have been discovered on Earth.

Evolution can begin practically anywhere, in interstellar space or on planets. But only on planets has it a chance of developing beyond chemical evolution. However, the ecospheres of stars—their life-supporting zones—have hitherto been greatly overestimated. Stephen H. Dole estimates in his book *Habitable Planets for Man* (see page 32) that the Earth would still support life if its mean distance from the Sun had been 30 per cent greater or 30 per cent smaller. This gave the Sun an ecosphere bounded on one side by the orbit of Venus, and extending on the other side to a point halfway between Earth and Mars. The error in these estimates was that correct allowance had not been made for the secular development of the Earth's atmosphere, the corresponding chemistry of the Earth's surface, and the accompanying slow increase in solar radiation. This was apparently done for the first time by Michael H. Hart in the summer of 1977 (see page 28). His computer simulation showed that the Earth's orbit could at the most have deviated by 5 per cent towards the Sun and by 1 per cent outwards from its actual position, and still have just permitted life to evolve. This reduces the Sun's ecosphere to one tenth of its previously accepted extent. Outside this strikingly narrow zone the consequence would either have been eternal glaciation as on Mars, or a killing, overheated Turkish bath, as on Venus. The negative results of the Viking expedition—so far as the discovery of organic material is concerned—are consistent with Hart's conclusions. Now that the Viking robots have failed to discover even traces of microorganisms on Mars, life is no longer anticipated on any other planet in the Solar System.

The evolution of *more complex* life-forms, the result of biological evolution, seems to be confined exclusively to planets. But for this it is not enough for the planets to be born within the narrow ecosphere of a star. They must also have the right mass—neither too large nor too small—otherwise either the surface gravity is too high, or the planet cannot retain a suitable atmosphere. The atmosphere must develop a sufficiently thick ozone layer to ward off the life-destroying ultraviolet radiation of the star; and it requires a high carbon dioxide content, to develop a moderate greenhouse effect and so prevent the seas from freezing. Further, the planet must turn on its axis, at the slowest, in 4 terrestrial days, or it will be too hot by day and too cold by night. It must also travel around its sun in an almost circular orbit, otherwise the seasonal changes will be too great. And even this special orbit must undergo only minimal variations over the millennia.

As a result of the researches of James D. Hays and other geologists (see page 34) it has been considered established that the ice ages were triggered by small and regular changes in the Earth's orbit. The current trend is in the direction of lower temperatures, with a new ice age in perhaps a few tens of thousands of years.

Not only the planet, but its sun too, must be of a special kind if it is to benefit the evolution of life: only stars assigned by astronomers to the spectral class G can be considered. (The Sun is of this class.) Larger and hotter stars have lifetimes which are too short to support a planet for 3000 million years, the minimum requirement for biological evolution. Smaller and cooler stars are ruled out simply on account of their temperature.

Hart's model studies brought together for the first time all the important factors of early terrestrial chemistry, primitive vulcanism, biochemistry and the gradual increase in solar radiation, and calculated their reciprocal influences as time passed. Although it may still be possible to refine them, it is already provisionally established that the regions around stars in which more complex life can develop are *considerably* smaller than previously assumed. It is possible that different physical conditions could have opened up other roads to the development of life, but they would all have had to find a way through the eye of the needle between icy deserts and greenhouse hells. Consequently, among the planetary systems of the Galaxy, far fewer planets will move in the ecosphere of a G star, perhaps by a factor of 1000. In 1970, Dole was estimating that almost 1 in every 200 of the 100 000 million (10^{11}) stars in the Galaxy was orbited by a habitable planet, leading to a total count of 645 million possible planetary homes. With the limits to the ecosphere now reduced, this number falls to less than a million habitable planets in our Galaxy.

Chemical evolution will probably take place on most of these planets. Whether the spontaneous appearance of life is always 'guaranteed' remains for the moment an open question. The astronomers Carl Sagan and Frank Drake, and also the molecular biologists Francis Crick and Leslie Orgel believe that life will always appear if it *can* appear. Mars turned out to be a great disappointment for these scientists. They and other exobiologists had believed that the origin of simple life was decidedly likely there. Nonetheless, molecular biologists still hold to the proposition that when life can appear it will. But it is evident that in the absence of precise knowledge of the necessary preconditions the limits for this have also been set too wide. The condition for 'can' may be defined by the work of the biophysicist Manfred Eigen: the precise requirement for the origin of life is that macromolecules produced by chemical evolution are able to organize themselves over a so-called hypercycle, and interact with one another (see page 44).

In hypercycles, proteins and nucleic acids work together in a cyclical way to increase their information storage capacity. When and in what physical and chemical conditions a given hypercycle sets in still requires considerable experimental investigation. Certainly an aqueous solution is necessary, a 'primeval soup' with a sufficiently high concentration of organic molecules and an adequate supply of energy. Of course we cannot yet be completely sure, but it seems that there never have been environmental conditions on Mars and Venus which would have permitted a hypercyclic coupling of organic molecules.

As soon as life is actually able to arise, it is at first chance which decides the basis of the genetic alphabet. After this, however, all alternatives are eliminated by an outright all-or-nothing selection process, and the unique accident is 'frozen-in' for all time. On this choice, further evolution is built up not at random but following a necessary pattern. Further evolution is not left to combinations arising through the play of chance, as the French geneticist Jacques Monod thought. This belief can arise if the end-products of evolution—biological molecules, genes and cells—are regarded only as alternative results of fortuitous combinations. Bacteria encode their genetic information with fully four million molecular symbols—the precise sequence of symbols is again only one out of the $10^{2\,000\,000}$ alternative sequences *possible in combination*.

As summarized by Eigen's colleague in Göttingen, Bernd Küppers:

> The fact that life is *a priori* an extremely improbable state of matter caused Monod to see the origin of life *a posteriori* as an absolutely unique lucky strike in Nature's 'lottery'. Monod was further convinced that a philosophy of existence based on molecular biology could be constructed on this insight. But Monod's ideas have been corrected on some essential points by the molecular theory of evolution, developed in particular by Eigen. Eigen's theory describes the evolution of life not as a purely chance event, but as a 'learning process of matter, in which the part played by chance is limited by the complementary nature of regular relationships'.[90]

But even when primitive organisms have arisen on a planet, evolution to more complex life-forms, and in particular the development of intelligence, seems by no means always to go smoothly. It can also encounter difficulties in this third state of the origin of life, biological evolution.

It is often assumed that after the appearance of life *all* planetary evolution will lead to intelligent life with 100 per cent probability.[61] This assumption rests on the belief that intelligence will increase the selective value of a species, so that mutants of greater intelligence will always prevail in evolution. Thus the development of intelligence would be inevitable. But the environment may well set limits to evolution before intelligence has developed. Evolution begins in interstellar space on its simplest level, that of chemistry. But it is too cold, there is too little incident energy, and there is no liquid in which organic molecules could encounter one another. Although surprisingly complex chains of molecules thrive in these precarious conditions, further complexity is frustrated. And at the next stage too, that of biological evolution, there are hurdles to be surmounted. On Earth, long before the beginnings of sexual reproduction, there was a critical phase—the transition from unicellular organisms to more complex organisms with many cells. This process took almost 3000 million years, the most protracted step between the primeval soup and the industrial revolution. In this interval oxygen-breathing organisms supplanted the previously dominant methane bacteria which consumed carbon dioxide, along with their relatives; and a 'reducing' atmosphere was transformed into an 'oxidizing' one by photosynthesis. This transition was perhaps the toughest biological barrier on the road to intelligence, on which evolution most often founders, and may come to a halt. Probably the oxygen burning of foodstuffs served the energy requirements of multicellular organisms better than the conversion of carbon dioxide into methane. 'It is possible that life can originate with relative ease in a planet with a reducing atmosphere and liquid water on its surface. The change however, into an oxidizing atmosphere might represent a very delicate transition which occurs only in a few special cases where all the contributing factors are exactly right,' says Michael D. Papagiannis.[77] He thinks that this transition succeeds on only one planet in a million.

The remainder of my criticism applies to the estimates of the number of technically advanced civilizations (existing simultaneously with us). The usual 'optimistic' view is that, the development of intelligence being inevitable, we can be confident that further evolution to a technological society capable of establishing communication would still be achieved by 1 intelligent species in 10. Thus there is little doubt that technological civilizations are numerous. Although some authors (Walter R. Fuchs and Günter Paul) are reluctant to give explicit statements on numbers, others (Stanley Miller and Leslie Orgel) consider all estimates 'equally (un-)reliable'. But most count civilizations capable of communication in the millions. The American space-philosopher Luis E. Navia is a modest exception with a 'cautiously estimated' 8333 (!) technological civilizations.[91] To Joachim Herrmann about 100 000 are a 'reasonable value, fixed neither too pessimistically nor too optimistically'.[92] Most writers regard one or more million as arguable (Ronald Bracewell, Frank Drake, Sebastian von Hoerner, Bernard Oliver[93], Carl Sagan and David Cameron). According to Sagan at least 100 of these civilizations should already be old enough for their signals to have been able to reach the Earth. Only Harlow Shapley outdoes them all in thinking—admittedly, as long ago as 1958—that the Galaxy is inhabited by 100 million civilizations capable of communication.

All these speculations must be judged in the light of the astronomical knowledge available today, and also of the biochemical findings already mentioned. Added to this, we must evaluate the astronomical observations of the past 20 years. Since 1960 there have been systematic searches for extraterrestrial civilizations—sporadic at first, then with increasingly large scope: with radio telescopes, with satellites outside the Earth's atmosphere as a branch of high-energy astronomy (X rays and gamma rays), and also with infrared and ultraviolet telescopes. The targets were Sun-like stars within the Galaxy. There have, in addition, been more and more searches of the whole sky, besides which a lookout has been kept for supercivilizations in our Galaxy and in some neighbouring galaxies. The result (so far as the discovery of extraterrestrial technology is concerned) has been negative—though a series of remarkable objects have been brought to light: X-ray stars, neutron stars as pulsars, quasars and black holes (and some of these sources were at first thought to be artificial!). For example, the radio point source discovered in the summer of 1977 at the centre of the Galaxy, no larger than half the Solar System, is very probably a gigantic black hole with a solar mass of several million. And the search for artificial infrared stars—anticipated signs of the activities of a supercivilization— has produced no results in 5 years. Instead the astronomers found heat-radiating dust clouds—stars in the process of formation. The result, as far as astronomy is concerned, is that the search for extraterrestrial civilizations is growing less and less speculative and becoming more and more part and parcel of the complete programme to explore the Universe. My *thesis* is that we are the only technologically developed civilization in the Galaxy. There are probably no supercivilizations which can rebuild whole galaxies either in the Milky Way or in its neighbours of the Local Group.

But the uniqueness of terrestrial civilization is not to be ascribed, as by Monod, to the improbability of the accident by which complex genetic molecules are formed. (These arise of necessity through the appropriate self-organization of the molecules.) Rather it is maintained that suitable planets, where evolution can overcome all biological barriers, are extremely rare. It is still conceivable that life in all possible stages of development, even endowed with 'intelligence', exists on planets: but its evolution may never have taken the path to technology, at least not up to the point at which telescopes are built, radio messages sent into space, planets dismembered and stellar energy tapped.

The relationship between civilization and technology, however, has not yet been made clear. Native races in the Amazon, in Africa and in the South Seas seem to have lived for thousands of years at the same stage of technology. Freeman J. Dyson considers the origin of intelligence and technology to be two separate phenomena: 'I make a sharp distinction between intelligence and technology. It is easy to imagine a highly intelligent society with no particular interest in technology.'[69] We may question this; technology is also a skill for survival. Unfortunately, the very difficulties of survival described above make it practically certain that other creatures will not have so pronounced an urge for self-preservation as we have. It cannot be ruled out as a possibility that 'at the same time' (that is, within a few thousand years) some few comparable civilizations have arisen in the Galaxy, but so far away that their signals have not yet been able to reach us.

Many people may counter this thesis with the objection that the astronomical data are still not clear enough. Indeed, so long as their statistical significance cannot at least be rendered plausible, their hitherto negative findings cannot be conclusively interpreted. In this connection the absence of all indications must not be over-interpreted. But if we take seriously the arguments about the physical limitations of habitable planets and the biological barriers, it seems to me that a much higher weight must be attached to the negative results presented by astronomy than has hitherto been assumed. The search programme of the 1980s will without doubt strengthen the significance of this conclusion considerably.

If the urge of life towards technology and then towards space travel were indeed so common and so inevitable, the whole Galaxy would actually be overflowing with technological intelligences, and the Solar System would long ago have been colonized and inhabited by them. The obvious *absence* of these phenomena suggests once again that mankind is unique as regards its technological stage of development. If this is so, a further conclusion suggests itself: intelligent life, perhaps rare but still possible, will hardly ever expand into all the surrounding space available to it. Although the development of diversity initially requires space, the cosmic absence of technological intelligence suggests that evolution will later take another direction—not towards further growth and expansion in space at any price, but perhaps towards further development inwards. This places us in a cosmically unique position. Together with responsibility for the future of mankind, we may also carry the responsibility of being the only highly technical life-form in the Galaxy. We must bear in mind that we are responsible for the whole Universe, since in a certain sense we are its peak. Many popular speculations have been developed to 'explain' the absence of any signs of extraterrestrial life: the zoo hypothesis—'we are being watched'—and the idea of the altered interests of a culture. In the zoo hypothesis the Earth is being observed or protected like a rare type of chimpanzee in our zoos. This theory—so constructed as to defy verification—is very close to the beliefs of the UFO watchers. It is as plausible as the idea of a divine daemon continuously supplying mankind with sense-impressions which only counterfeit a world and a Cosmos for us.

But if a watch by spaceships—perhaps hidden in the asteroid belt—is proposed, we can still test it by going there directly. The idea is also widely held that civilizations developed to a much higher level will no longer care to establish contact with us. We examine microorganisms, but normally we do not communicate with them.[84] But we would if we could! More students of communication than ever before are engaged in years-long experiments to establish some understanding with dolphins and chimpanzees. The other drawback to the idea of explaining the absence of extraterrestrials by changing interests is that *any* civilization which has found its way to the possibility of interstellar travel and communication would have to have abandoned this achievement again. This again is most implausible.

In my opinion the chances of coming upon traces of or signals from extraterrestrial civilizations are very slim.

If the search for extraterrestrial intelligence were to depart too much from modern astronomy, I should consider this search a waste of time and money. Correspondingly, less money should be spent on *sending* messages. But a search programme, practically identical with the study of the new types of waves discovered by astronomy (infrared, ultraviolet, X-ray, gamma, neutrino and soon perhaps even gravitational waves) seems justified. For only by the analysis of all accessible channels of information in the study of the heavens will we realistically be in a position—should we actually, against all expectations, come upon traces of another civilization—to identify it with certainty, and not succumb to our own wishful thinking.

part III
The conquest of space

The Galactic Club

Many people, concerned about the possibility of invasion by extraterrestrial civilizations, have questioned whether we should broadcast signals at all, so as to give away our position and betray our solar hiding-place. The astronomer Zdeněk Kopal of Manchester University exclaims: 'For God's sake let us not answer!' In case anyone is still disturbed about this—it is already far too late! Not simply because a short message was transmitted in 1974 with the Arecibo telescope: this largely symbolic message still has a good chance of never being received. But all the other ordinary radio signals from terrestrial broadcasting can already be received at interstellar distances, so long as they are more than 1 per cent stronger than the radio noise from the Sun. In the last two decades more than 1000 such signals have left the Earth every second with the speed of light. Every year the spreading wave front takes in about 20 more stars. And as if that were not enough, the first interstellar probes are already on their way.

The first artificial, man-made celestial body will leave the Solar System in 1987. The American space probe Pioneer 10, launched on 3 March 1972 (followed by Pioneer 11, launched 4 June 1973) was the first satellite to cross the asteroid belt and pass by the planet Jupiter, taking 300 close-up pictures and investigating its magnetic field. The crossing of the asteroid belt incidentally proved less hazardous than had at first been feared.

In 1979 Pioneer 10 crossed the orbit of Uranus, and 9 years later it will move into interstellar space with a residual velocity of over 15 kilometres a second. Although a rocket can escape the gravitational field of the Earth with this velocity, the probe moves along at something of a snail's pace. It takes almost 18 000 years to travel the distance of 1 light-year. It is heading between the constellations Orion and Taurus, in a region fairly empty of stars, towards the edge of the Galaxy. It is thus hardly ever likely to be picked up. In order to escape the gravitational pull of the Solar System, Pioneer 10 was accelerated on 3 December 1974 by a close approach to Jupiter. Pioneer 10 was the first spacecraft to be accelerated by the gravitational field of a planet—the principle of a 'gravitational slingshot' (see 'Binary stars as gravitational catapults', page 251).

Both Pioneer 10 and 11 carry a plaque—see Figure 56—designed by Carl Sagan and Frank Drake (and drawn by Sagan's wife Linda). The plaques, made of especially durable gold-plated aluminium, provide information about the position of the Earth and the time of the probes' production and launching—rather like a cosmic picture postcard. The periods of the 14 pulsars nearest to the Earth are indicated in binary code. Pulsars—neutron stars which emit radio signals like lighthouses as they rapidly rotate—are cosmic calendars: their rotation decreases slowly at a rate which is constant over long intervals. If someone finds the probe and identifies the pulsars engraved on the plaque, he can determine, from the

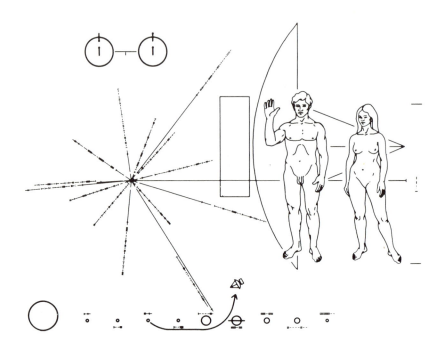

Figure 56 Gold-plated plaques on board the space probes Pioneer 10 and 11, launched towards Jupiter, will leave the Solar System at the close of the 1980s as the first material message from mankind. At the suggestion of astronomers Frank Drake and Carl Sagan, information for possible extraterrestrial beings was engraved on the 15 × 22.5 cm plaques: the Earth's position relative to 14 pulsars (whose pulse periods are given in binary code); a man and a woman in front of a representation of the probe; the situation of the Earth relative to the other planets of the Solar System and the path taken by the probe; together with a hydrogen molecule at the upper edge, to provide a scale.

difference between the rotation periods observed by him and those given on the plaque, how long the probe has been on its journey. The route taken by the probe through the Solar System is engraved at the bottom of the plaque, showing it leaving the third planet and swinging across near the fifth planet, Jupiter. On the right, in front of the profile of the space probe, two human beings are pictured, a man and a woman, with 'pan-racial characteristics'; the man is waving at the reader. A hydrogen molecule is shown for scale at the top of the picture. From the frequency of 1.42 gigahertz and the wavelength of 21 centimetres at which the hydrogen molecule radiates, the data on time and length can be calculated. It is not impossible that Pioneer 10 and 11 will still be pursuing their lonely way through the interstellar spaces of the Galaxy when the Earth has been swallowed up and incinerated by the changing, expanding Sun in its red giant stage.

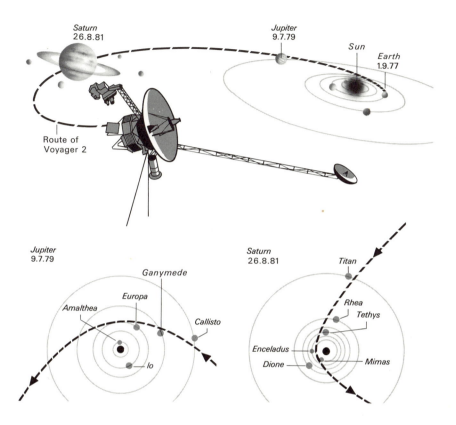

Figure 57 Route of the space probe Voyager 2. The launches of Voyager 1 and 2 took place on 5 September and 20 August 1977 respectively. In July 1979 Voyager 2 took close-up photographs of Jupiter and its four largest moons—Callisto, Ganymede, Europa and Io—as well as Amalthea, the innermost moon. Voyager 1— not shown here—took a more direct route, passing Jupiter despite its late start by the end of March. By November 1980, it had successfully surveyed Saturn and its rings. Four years after launching, on 26 August 1981, Voyager 2 also skimmed close by the outermost ring of Saturn. It now flies on to Uranus, where it will arrive in 1986, finally passing Neptune in 1989.

The two Pioneer probes seem to have started a fashion for sending messages into space. On 20 August and 5 September 1977, NASA launched the spacecraft Voyager 2 and 1 (in this order) with the rocket combination Titan III E-Centaur (see Figure 57). Among other things, their television cameras were set to examine the mystery of the rings of Saturn during the expeditions, first in late 1980 then again in August 1981. The long-standing astronomical controversy over the rings—whether they consist of icy fragments or rocks—has, as a result of the Voyager data, finally been settled. The craft have certainly furnished us with beautiful and detailed photographs of Jupiter and its moons and Saturn (see Figures 20, 21 and 22), and later the planets Uranus and Neptune will come under scrutiny. A flyby of Uranus is calculated for January 1986, and Voyager 2 will reach Neptune, 4 light-hours away, in 1989, 12 years after leaving the Earth. NASA scientists hope to maintain contact with the probes with the help of their Deep Space Network for a total of 33 years, long after they have crossed the orbit of Pluto, the last planet before interstellar space.[1]

The Voyager programme was and is concerned with the outer planets, still largely unknown, and especially Jupiter. Pictures of its mysterious Red Spot, assumed to be an eternal whirlpool in the hydrogen–helium atmosphere, were transmitted to the Earth with 100 times the resolution that Pioneer 10 and 11 were able to achieve in their flyby in March 1974.

The space vehicles, almost a ton in weight, carry along with their 10 scientific instruments an unusual cargo: a sound and video recording. It was especially produced for the Voyager mission on a practically indestructible disc of copper. Equipment for playing it and instructions are thoughtfully provided. The content of this offering, which like the Pioneer plaques was suggested by Carl Sagan, is a blend of the Earthliest of the Earthly: more than 100 photographs, including landscapes, falling leaves, children, the scientist Jane Goodall with her chimpanzees, and a human couple engaged in sexual intercourse; tape recordings of words of greeting spoken in 55 languages, some of them extinct like Hittite and Sumerian, but also English, German, Russian and Chinese; a pot-pourri of music from Bach to Beethoven (the Fifth Symphony!), folk music, but also rock and soul music by Chuck Berry and Louis Armstrong. Then the noises of the Earth: volcanoes, whales, frogs, a truck, and the launching of a Saturn V rocket. The crowning touch is given by the words of Jimmy Carter: 'We are trying to survive our own time . . . to take our place in a galactic community.'[2]

Even if the sending of such messages may be more like a cosmic joke, they demonstrate a possibility, even while we are still opening up our planetary system: robot stations can be dispatched to other parts of the Milky Way. The radioastronomer Ronald N. Bracewell is among those who support the idea that a Galactic Club of advanced creatures, perhaps long in existence, might make use of interstellar probes for communication as well as to observe potential future club members.[3,4] Although radio waves are certainly the best medium of communication in certain conditions, there are situations in which space probes, by complementing radio techniques, might facilitate the discovery of other societies and bring success more quickly. Clearly space probes, along with personal visits by spaceship, offer a direct means of discovering an inhabited planet on which there is not yet any radiotechnology. But in other ways radio contact is of variable utility, depending on the distance to the nearest civilization. For short distances (Case I in Table 12) Bracewell considers radio methods certainly the most favourable, making discovery and information exchange possible most quickly. Bracewell calls this the Ozma case. As we mentioned, the aim of the Ozma project was to listen to the two nearest Sun-like stars, Epsilon Eridani and Tau Ceti, for artificial radio signals. But undertakings of this sort only hold promise of success if the population of the Galaxy is so large that civilizations are separated on the average by less than 30 light-years. Even Bracewell himself does not believe this.

But the greater the distance to the putative neighbour within the Galaxy, the more stars pregnant with life appear on radioastronomers' checklists. Within a sphere of 100 light-years radius this already amounts to about 1000 stars like the Sun. Accordingly, there is less observing time available for each individual candidate. During the Ozma II project carried out by the radioastronomers Ben Zuckerman and Patrick Palmer (see page 159), each of the nearby Sun-like stars—about 650, at a maximum distance of 75 light-years—was under observation for only 30 minutes. (However, some 'suspect' stars were re-observed.) Even though it is often assumed that an older society, superior to us, will take the initiative (or will already have taken it), we are faced with a choice from among 1000 directions.

Directional reception guarantees the greatest range (see Part II: 'Strategies for distance, direction and frequency', page 142). If, however, the sender has the same idea, he has a choice of 1000 possibilities when beaming his transmitting aerial. Beyond 100 light-years the choice can no longer be narrowed down solely on the basis of the astronomical typology of the stars, at least for civilizations of Type I. From a great distance the Earth looks essentially the same today as it did 1000 years ago. And even if radio signals had reached the Earth as recently as 100 years ago, nobody could have received them. And even if we were to send an answer to-day to an interstellar message received yesterday, it would be lost if, in 100 years' time, the addressee were just examining one of the other stars for an answer.

First contact by space probe?

In Part II we discussed in detail the strategies by which these difficulties in making contact could be avoided. But Bracewell has another suggestion which he believes will prove less expensive to the sponsors and which is also politically acceptable as a long-term scientific project. By his plan it will be rather a long time before contact is established. But Bracewell thinks that radio searches may also require a large expenditure of time (though the radio connection would then be all the more reliable).

Interstellar space probes would be sent off to interesting stars within a few hundred light-years (see Table 12). This first phase, the sending of the probes, would represent the most wearisome part of the search for contact. They would certainly be faster than Pioneer 10, but they would still take, on average, 1000 years to reach their destinations. After this waiting period, however, things might go forward with fewer problems. As soon as a space probe approaches its target

Table 12 Search and communication by radio or probes?*

	Technical civilizations in the Galaxy			Special cases	
Case	I	II	III	IV	V
Frequency	Very numerous (more than 50 million)	Less numerous (a few million)	Very rare (less than 100 000)	We are alone: in the Galaxy	in the Universe
Mean distance	Less than 30 light-years	30–300 light-years	300–100 000 light-years	More than 1 million light-years	—
Search	Radio	Probes	?	Radio	
Communi- cation	Radio	Radio	?	One way only	

* After Bracewell.[3]

star it fires off its last rocket, brakes and swings into an orbit as nearly as possible within or near the habitable zone in the equatorial plane of the star, where the planets are situated. With the aid of the specially pre-programmed instruments it carries, the probe can guide itself into this position unaided.

Once the space probe has established itself in a suitable orbit near the inhabited planet (if there is one) the operation proper begins. Solar cells provide the energy: this is in any case the most economical method. Even if the probe does not encounter any intelligent life, it can still collect and transmit back to the home planet physical, chemical and biological data about the planets which are there (number and size, their atmospheres, existence of primitive life-forms). Thus, for instance, the colonization of other planets can be prepared for—a possible side-effect of Bracewell's stellar scouts, independent of the search for contact. If the probe discovers intelligent life it can immediately send home the news that the neighbours sought are here. It is conceivable that this has already happened in the Solar System. Perhaps a tiny, inconspicuous satellite is circling somewhere between Venus and Mars, and for many years now important news has been on its way to another star. And then? It would not be economical if the mini-robot continued to circle among the planets dumb and inactive. The probe could now devote itself to the Earth, make its presence known to us Earthlings, provide information about its makers and invite communication. Bracewell identifies three phases in the process: the probe must attract our attention, tell us the identity of its star of origin, and set up a code and transmission schedule for communication back to base.[3] Bracewell has already suggested the way in which the probe could bring itself into our field of view as conspicuously as possible: the best way would be by *repetition* of specific radio news bulletins intercepted by the probe. If one evening the radio announcer says: 'That is the end of our programmes for tonight. A very good night to you all,' and 5 minutes later, just before we turn off the radio, we hear, like an echo: 'A very good night to you all,' it will cause notice. Step by step, as in a telephone conversation in which each participant must first ascertain the language of the other, a common code could be constructed. The first message sent to the home planet of the probe, drawn up after an exchange with it, would then be a reliably addressed (beamed) message, perhaps comprehensible to the receiver without further effort.

Bracewell's bridge of contact by probes avoids the difficulties of a search by radio soundings and eliminates the other problem of interstellar communication: that of launching one-sided messages into the (relatively) unknown, without the opportunity of feedback which might clear away the initial obstacles to understanding. This immediate feedback is provided by the probes in much the same way as the computer program from Andromeda in Fred Hoyle's novel (see Part II: 'A computer program from Andromeda', page 114). But it is not quite clear why Bracewell thinks the dispatching of probes financially cheaper than for instance a Cyclops programme keeping 1000 stars under constant observation. Bracewell's probes would be considerably more expensive than space robots of the Voyager type. Their communication equipment must operate over a distance of several hundred light-years, not to mention the fuel requirements and the rockets necessary for navigation within the alien star system. And if about four probes are successfully launched every year, it will still take 250 years to send off 1000 probes.

Besides this, the probe strategy only promises success if the conditions for Case II in Table 12 are fulfilled: that the Galaxy is dotted with 10 million technological cultures. If there are fewer, as in Cases III and IV, the probe programme becomes rather pointless. Beyond the limit of 100 light-years, single Sun-like stars can no longer be properly distinguished optically. Moreover, the stellar robots would simply require too much time even for a patient civilization: at least 10 000 years. In view of the arguments of Part I and II (see in particular the conclusion to Part II, 'We are (still) alone', page 222), Bracewell's belief in 10 million galactic cultures capable of communication is not tenable. Bracewell has advocated his probe idea since 1960 (in a lecture at Green Bank, where Ozma was carried out) after reading the article by Cocconi and Morrison.[5] It is rather surprising that he has not since then critically examined his premises (Case I) to which he himself attached the greatest weight.

Interstellar travel—journeys of no return

Space probes, followed by spaceships—today this seems at least an attainable Utopia. Too slow and too costly for a lonely human race, but already possible with present-day technology to a financially able, emigration-minded minority, the (gradual) colonization of the Universe may be carried out. For travel purposes, however, space journeys will remain a local enterprise: only the Solar System is on offer to trippers. Those bold enough to go further do so with no return ticket; interstellar space travels are one-way journeys.

The reason for this is the enormous distance separating the stars and the correspondingly long travelling times. Even if a spaceship forges ahead with 10 per cent of the velocity of light—about 1000 million kilometres an hour—a trip to Sirius still takes 88 years. This clearly presents a problem for short-lived creatures like man. This might be circumvented by the technique of 'suspended animation', in which the metabolic functions are reduced to practically zero. The ageing process would be correspondingly slowed down. It is true that we have not yet mastered the art of 'freezing' warm-blooded animals and then thawing them out. But biologists are investigating how hibernating animals succeed in surviving in a deep sleep for months without nourishment. There are no grounds for assuming that intelligent extraterrestrials will have life-spans similar to our own. For longer-living individuals of other societies, an interstellar journey may not mean the squandering of almost a whole life, but perhaps only a welcome change.

Many suggestions have been made to overcome these problems of space travel. For example: by making use of relativistic time dilation, the effect by which, in a spaceship moving relative to the Earth with a velocity near that of light, the time on board passes 'more slowly' than for those who stay behind. This would in fact shorten the travel time considerably for space travellers. But let us assume a convenient journey under physical circumstances like those on Earth: for the first half of the trip a constant acceleration of $1g$, the Earth's surface gravity, and for the second half an equally strong deceleration. On these terms, the crew would reach the nearby star, Epsilon Eridani, in 5 years, and the open cluster of the Pleiades, 410 light-years away, in 10 years of spaceship time.[6] But after the return journey (in similar conditions) the space travellers might not recognize the Earth—too much Earth time would have elapsed. After a round trip to a destination 460 light-years away, which takes 25 years for the crew, mankind would have aged by 910 years.

The snag to these simple calculations (which are based on Einstein's Special Theory of Relativity) and the space fantasies which have been spun with this yarn is the enormous expenditure of energy involved which would in practice make it impossible for mankind to undertake relativistic space travel. It would be difficult even for supercivilizations. At the beginning of the 1960s Sebastian von Hoerner[7] and Edward Purcell[8] estimated that, even in optimum conditions of nuclear propulsion, travel near the speed of light would require the expenditure of 3 megawatts per gram of payload. In addition there are the numerous transmission stations to beam the radiation in a particular direction and to propel the spaceship. To raise 10 tonnes of payload to within 2 per cent of the velocity of light in 2 or 3 years of spaceship time, von Hoerner calculated somewhat grotesque requirements: 'We would need 40 million annihilation power plants of 15 megawatts each, plus 6 billion [6×10^9] transmitting stations of 100 kilowatts each, altogether having no more mass than 10 tons. . . .'[7] If there are fewer propulsive units, perhaps 'only' 40 power stations and 6000 transmitting stations, 98 per cent of the velocity of light can only be attained after 2.3 million years.

So relativistic journeys are out. Travel speeds of 10 per cent of the velocity of light are more realistic. The ejection velocity of the propellant more or less determines the attainable limiting velocity. In chemically propelled systems the gases are emitted at 3 kilometres per second. Thus several rocket stages are needed in order to leave the Earth, for which a speed of 11.5 kilometres a second is necessary. For a flight to the Moon and back a chemical combustion system demands about five stages (including the Moon-lander). This demands at launching a rocket weighing 1000 times as much as the payload finally brought back. Thus chemical propulsion will only serve for space traffic in the region of the Earth and its immediate neighbours—the Moon, Venus and Mars.

Hydrogen bombs for rocket propulsion

Nuclear fuels, on the other hand, permit of propellant velocities up to 10 per cent of the velocity of light—30 000 kilometres per second—and therefore, in principle, a correspondingly greater travel speed. Since 1958 serious work has been carried out in the United States, under the name of Project Orion, with the aim of designing a spaceship which would be propelled by nuclear explosions. The idea came from Stanislaus Ulam (at Los Alamos); and the physicist Ted Taylor, who was previously responsible at Los Alamos for the development of nuclear weapons, led the project, details of which remain secret even today. Freeman J. Dyson, then a young man, was also in the group.[9]

The great advantage of the Orion spaceship is that for journeys within the Solar System the launching weight need be no more than 10 times the payload. Also, hydrogen bombs are the most effective means of burning the relatively cheap fuel deuterium, a heavy isotope of hydrogen. The precise final velocity Orion can achieve depends on the construction. What will such a hydrogen bomb engine look like? The simplest structure would be a hollow hemisphere 20 kilometres in diameter, at whose centre the bombs explode. The thrust is then transferred to the ship via a blast collector and shock absorbers (see Figure 58). Here a problem must be solved. The inner surface of the hemisphere must be able to withstand the heat wave and the pressure of the explosion. A metal layer only 1 millimetre thick would suffice for this. The structure must then take a shock of no more than a tenth of an atmosphere pressure for a hundredth of a second. Finally the heat of the explosion must be disposed of. Here copper suggests itself as a good conductor of heat. If a bomb is detonated at intervals of 100 seconds, there is enough time between blasts for cooling. If the risk of a certain amount of wear to the surface by heating is accepted and a uniform accelaration of 1g is desired, with an increase of velocity of the spaceship of 30 kilometres per second per explosion, one bomb can be detonated every 3 seconds. A diameter of 2 kilometres will suffice for the shield, which takes up about 10 per cent of the energy of the explosion. The shock absorbing layer would then be 75 metres thick. With an acceleration of 1g, cruising speed would be reached within 10 days.

However, Project Orion was taken no further than the testing of models with chemical explosives on a terrain in California. Finally, in 1965, it was completely halted and the Saturn and Apollo programmes made the running. But in 1973 the Orion idea was taken up again in the United Kingdom as Project Daedalus. The aim of this project is to study the feasibility of a robot-controlled unmanned spaceship to send to Barnard's Star, which may be orbited by a planet. Propulsion by nuclear bombs in the Orion style is expected to give the probe a speed as great as 50 000 kilometres a second. Daedalus would then reach its destination in 25 years. The study is at present continuing.[10]

In fact, the ideal fuel would be photons, or electromagnetic waves: they would leave the spaceship with the highest velocity physically possible, that of light. Photons could be produced in sufficient quantity in an engine in which matter and anti-matter were brought into contact. The flash of matter and anti-matter in mutual destruction would produce the desired energy-rich radiation. But the problem here lies in providing the material. For a round trip to a destination 100 light-years away, millions of tons of anti-matter would be needed. The technology which would be required to set a matter–anti-matter reaction going and to control it, not to mention the production and storage of the anti-matter, lies far beyond the foreseeable future.

Interstellar ramjets

A totally different idea for propelling a spaceship has been put forward by R. W. Bussard: the interstellar ramjet.[11] Its principle is very simple. A rocket with a collecting surface shaped like an inside-out umbrella in front, and an engine behind, travels through the space between the stars. The rocket uses the umbrella to collect interstellar gas (in particular, protons), burns it by fusion in thermonuclear reactions, and ejects the gas behind it again with increased velocity. Calculations have shown that, in principle, a rocket which does not carry its fuel with it from the start, but picks it up continuously on its way, transfers the kinetic energy of the combustion gases to the spaceship very advantageously. This, of course, only works if the spaceship encounters enough fuel—that is, if the interstellar medium is dense enough. This would be the case,

Table 13 Comparison of ramjet and conventional propulsion for interstellar journeys of different lengths*

Destination	Alpha Centauri (4.3 light-years)		Canopus (100 light-years)	
	Conventional rocket	Ramjet	Conventional rocket	Ramjet
Ratio of initial to final mass of rocket	10^4	10	10^4	10
Earth time for round trip	38 years	40 years	871 years	260 years
Ship time for round trip	37 years	38 years	847 years	130 years
Mean speed of ship as percentage of light speed	22.4	21.5	23	77

* After Roberts.[14]
The use of a (currently unrealizable) ramjet instead of a conventional rocket (technically feasible today) is unlikely to be of advantage on the relatively short trip to Alpha Centauri. Only on an expedition to a star 100 light-years away would the ship arrive at its goal noticeably faster.

for instance, in large hydrogen clouds. Propulsion by ramjets also essentially assumes that controlled nuclear fusion is possible. Despite numerous terrestrial experiments in this direction the result of these attempts cannot yet be predicted.

To collect the low-density gas between the stars in sufficient quantities, the collecting surface must have a diameter of at least 10 or 20 kilometres. Even if we ignore the problem of the construction of the collector, the protons and deuterium ions of the interstellar plasma must be directed to the reaction motor, for example by magnetic fields. Here we find another snag to Bussard's otherwise attractive idea. 'This is a horribly unstable situation,' says Philip Morrison, who has made careful studies of ramjets.[12] The charged particles will already emit radiation on capture, possibly just as much as can be obtained later from fusion.[13]

The other difficulty concerns the ability to handle the strong magnetic fields of the huge collecting screen. Perhaps one day, at the end of the next century, our science will have mastered the 'very formidable problems associated with ultra high-temperature reactors, the low-proton-proton fusion rate, and the drag induced by the large ramscoop [magnetic collecting screen]'.[14] Then, and only then, vehicles equipped with ramjets might prove themselves superior to all other spaceships. For relatively short interstellar journeys, such as to the nearby star Alpha Centauri, the efficiency of ramjet propulsion would be no better than that of other rockets. As Table 13 shows, however, the advantage would be quite considerable in a flight to the star Canopus, a distance of 100 light-years.

The laser-driven spaceship

In nuclear and chemical rockets which carry their propellant with them the transfer of thrust is extremely poor. The greater part of the kinetic energy liberated in combustion is carried away by the gases themselves, and only a small portion of it is transferred to the rocket. A better way to propel the rocket would be to provide it with a 'firm base', which will take up the momentum but not too much kinetic energy. Jules Verne had already grasped this principle, if only intuitively. In his *Journey to the Moon* (1865) the mooncraft is fired from the Earth by a huge cannon. (In fact, none of the passengers would survive the acceleration of this violent start.) Given the great mass of the Earth, the momentum imparted to the Earth by the shot would be negligible.

A modern version of Verne's propulsion principle seems to be theoretically within reach, however, with the aid of laser beams. This idea was first considered by the Hungarian scientist G. Marx in 1966.[15]

This is how the propulsion system would work (see Figure 58). The spaceship leaving the Earth has a mirror attached to it pointing towards the Earth. A beam of light is directed from Earth to this mirror, which reflects it. The momentum transferred from the reflected beam of light propels the craft forward. This principle has the advantage that practically the whole of the drive system remains on the Earth, thus avoiding a great deal of expense. Considerable momentum could be imparted to the spacecraft by an Earth-bound propulsion system, as with electrically driven jets. The disadvantage is the range: the distance over which energy can be transmitted is limited to the Earth–Moon system.

In the absence of laser techniques such a system long remained unthinkable. Radio waves and microwaves would be fanned out and scattered in the atmosphere to such a degree that at great distances they could no longer be concentrated on the rocket's receiving system. But 'with the advent of lasers we are in possession of a source of energy that can be transmitted through the atmosphere without great loss by dispersion or absorption,' writes Arthur Kantrowitz of the Avco-Everett Research Laboratory.[16] The cost would be reduced above all by the fact that the drive can be fed by current produced on the Earth. Kantrowitz again: 'Consider what it would cost to get one kilogram into Earth orbit, if we could connect it with an electric cable and supply it continuously with energy. The kinetic energy of an object in low Earth orbit amounts to about nine kilowatt hours per kilogram, which demands only a few cents' worth of electricity.'

For example, to catapult a vehicle away from Earth with an acceleration of $10g$, 10 times the Earth's gravity, around 400 megawatts would have to be expended per tonne of material. This would be enough to give an orbital velocity of 8 kilometres per second. Kantrowitz therefore considers an arrangement of ten 100-megawatt lasers to be a workable system. The laser energy could still be

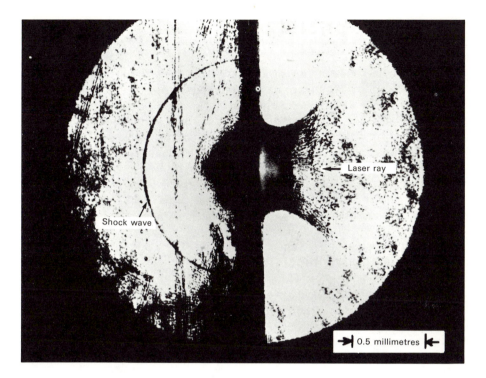

Figure 58 Preparatory experiments on laser fusion. The picture shows a short-exposure shadow projection of the 'impact' of a laser pulse on a layer of transparent plastic (enlarged 15 times). The laser pulse has arrived from the right; the material evaporated by the impact is spraying in the direction of the beam. Within the plastic sheet the pulse produces a hemispherical compression wave: it is proposed to use this shock wave to raise frozen hydrogen spherules to fusion temperatures. The exposure time of the photograph is 5 picoseconds (5×10^{-12} s); it was taken by the laser research group at the Max Planck Institute for Plasma Physics in Garching. (Courtesy: Max-Planck-Gesellschaft.)

concentrated well on a mirror 2.5 metres in diameter at a distance of 100 kilometres.

The futuristic note in this story lies in the required laser output. In continuous use, modern lasers yield only a few tens of watts as laser light, and to produce this the laser already uses several kilowatts of current. The efficiency of lasers is extremely low. Much higher output is achieved in short laser pulses, but only for picoseconds (10^{-12} seconds) at a time. We are still several orders of magnitude short of the desired continuous supply of 100 megawatts. Kantrowitz believes that this advance will be made: 'I do not know of any physical limitations to stop us from doing it.' He also believes it possible to master the techniques of accurately directing and tracking the laser beam, and to counter the possible reactions of the atmosphere to the intense laser radiation. The lasers will need to be fired without interruption for about 15 minutes from the time of launch until the rocket is in orbit.

The feasibility of laser propulsion has still to be demonstrated, and laser technology is still too young for its full potential to be estimated. But the Soviet astronomer N. S. Kardashev already dreams of a combined application (yet to be devised) of nuclear, laser and ramjet propulsion for interstellar travel to a neighbouring civilization. 'The space capsule will be accelerated to within a fraction of the velocity of light with nuclear fuel. Then an X-ray laser propels the craft and accelerates it to almost the velocity of light. Ramjet propulsion is only for use in navigating interstellar gas clouds. On arrival the craft can be braked in the same manner, with the active assistance of the receiving civilization. The return launch and arrival on Earth follow the same pattern.'

To the stars with laser fusion

For several years lasers have also been applied to enhance the production of energy by nuclear fusion. The simplest fusion process occurs when two nuclear particles of the proton type are compressed together so strongly that they fuse. In the process, helium nuclei are produced and energy is liberated; but the fusion reaction only gets going if pressure and temperature are about as high as in the interior of the Sun. As with any good reactor, a laser fusion reactor could be adapted for interstellar propulsion (Figure 59). Attempts to produce controlled fusion with the aid of powerful laser beams are being made, mostly in the United States and the Soviet Union. Here, deuterium and tritium, heavy isotopes of hydrogen, are used as starting materials. A deep-frozen pellet of this material, smaller than a millimetre, is bombarded with pulses of laser radiation until the temperature is high enough to set the fusion process going. This reaction temperature is reached at 100 million degrees. But heating alone is not enough: the thermonuclear reaction can only be effectively started if at the same time the little pellet is compressed to 10 000 times its normal density. It is therefore irradiated by lasers simultaneously on all sides, so that its whole surface is covered.

Figure 59 Rocket propulsion today and in the future. Bomb-driven spaceships seem at present the only realistic vehicles for interstellar space travel. Exploding atom bombs impart momentum to shock absorbers and thus to the payload of the ship. The shock absorbers are arranged, in the simplified model shown here, in a hemispherical shell. With this vehicle, travel speeds up to a tenth of that of light (30 000 kilometres per second) could be reached. The interstellar ramjet captures interstellar gas via a collecting surface of several thousand square kilometres and directs it into a reaction chamber. Owing to the instability of the gas flow and radiation losses, this type of propulsion will not be practicable within the foreseeable future. Besides laser fusion and propulsion by laser radiation, the acceleration of spaceships by binary stars is also conceivable in principle.

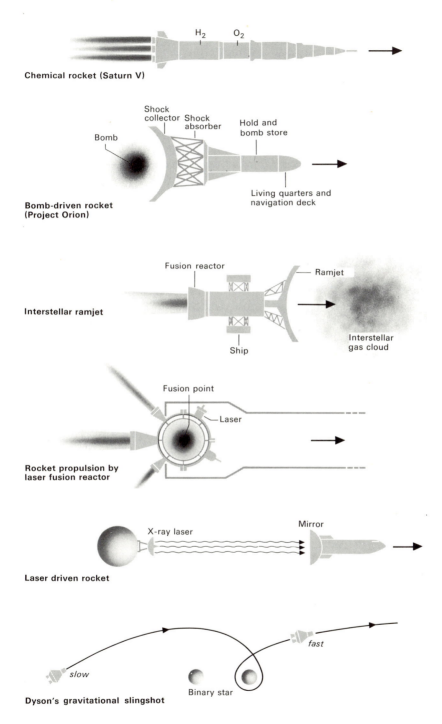

Chemical rocket (Saturn V)

H₂   O₂

Bomb-driven rocket (Project Orion)

Bomb

Shock collector   Shock absorber

Hold and bomb store

Living quarters and navigation deck

Interstellar ramjet

Fusion reactor

Ramjet

Ship

Interstellar gas cloud

Rocket propulsion by laser fusion reactor

Fusion point

Laser

Laser driven rocket

X-ray laser

Mirror

Dyson's gravitational slingshot

slow

fast

Binary star

The energy must be 'shot' at the pellet in a pulse of a thousand-millionth (10^{-9}) of a second's duration, so that its outer layer evaporates almost explosively. The interior of the pellet is thus compressed by implosion with a velocity of more than 50 kilometres a second, until the rising back-pressure stops the process. But by now pressure, density and temperature are high enough to kindle the reaction.

At present the drawback to laser fusion—as with the propulsion of rockets from the Earth's surface by lasers—is that lasers are still too weak. Gas lasers powered by carbon dioxide offer the best prospects of developing powerful sources of radiation with lasers. They convert the energy of the current fed into them into the optical energy of the light beam with the best relative efficiency at present attainable, at a rating of 10 per cent—ordinary gas lasers reach only 0.1 per cent. But lasers using other gases, the noble gases argon, xenon and krypton, or metallic vapours of copper, tungsten or manganese, are also under consideration.[17,18]

To adapt a fusion reactor for rocket propulsion, microfusions must be kindled in rapid succession. As currently envisaged, the liquid deuterium and tritium are first pumped from a tank and converted into deep-frozen pellets. An injector—a kind of rotating catapult—then shoots the pellets into the propulsion chamber with a velocity of 3 kilometres per second, at a rate of about 500 every second. Here the lasers and the impinging pellets are precisely aligned with one another. The lasers fire a pulse every two-thousandths of a second, each time a pellet goes through the fusion point. The charged particles sprayed out by explosions are led out of the reaction chamber by strong magnetic fields in the direction required by the course to be steered. This provides the spaceship's fusion drive.

Laser experts are optimistic about the attainability of their goals, but whether their optimism is well founded remains to be seen. Certainly work on fusion is going ahead at full speed, and of the two routes that have been taken—constraint by toroidal magnetic fields and implosion heating by laser bombardment—many fusion experts see greater promise in the second.

Recently Gerald O'Neill, a high-energy physicist at Princeton, put forward another suggestion for propulsion which, like laser propulsion, may be especially applicable to transport problems within the Earth–Moon system. In essence, ballast is to be expelled during the flight not by ejection, as has been done hitherto with used rocket stages, but using a 'linear accelerator'. Well known in particle physics, this is a machine for accelerating charged atomic particles to high speeds using magnetic fields produced by superconducting coils. In the rocket application the ballast, say empty propellant tanks, would first be reduced to suitable fragments, then accelerated away at high speeds by this method. The idea has already been tested in practice—a model 2 metres in size, constructed by a group at the Massachusetts Institute of Technology, was successfully tested in 1977. It moved objects from 0 to 130 kilometres per second within a tenth of a second, corresponding to an acceleration of $35g$. A system suited to space travel is now to be developed in collaboration with NASA, and built in Earth orbit in the 1980s together with the Space Shuttle. As an interim goal of his research programme O'Neill described in 1977 'a vacuum synchrotron motor ten metres across . . . designed for an acceleration of $1000g$ and with a terminal velocity of 1000 kilometres per second'.

Although this method of propulsion is still new and little tested, it does appear promising. Other ideas with which O'Neill has been associated have been very successful—back in the 1960s he made considerable contributions to the development of the 'double storage ring' for particle acceleration. In all experiments up to then the streams of accelerated particles had been directed to a stationary target. O'Neill's idea was to collect the accelerated particles in a magnetic ring in which the collison was to take place. A second stream of particles was accelerated in the opposite direction and aimed towards the first. This dramatically increased the energy obtainable from particle collisions. O'Neill's work of 1959 already contained the idea of proton storage used in the particle accelerator CERN near Geneva, with which psi particles and the existence of 'charm' were discovered. O'Neill's very solid plans for space colonies will be discussed later.

Binary stars as gravitational catapults

When the American space probe Pioneer 10 passed within 131 400 kilometres of the giant planet Jupiter on 12 April 1973, it came away with not only a multitude of pictures and scientific data, but something else as well: gravitational energy. More precisely, it drew energy from the gravitational potential between the Sun and Jupiter, which has 318 times the Earth's mass. After its flyby, Pioneer 10 shot with increased velocity towards its new target: the open space beyond the Solar System. At the same time, Jupiter and the Sun drew marginally closer together on account of the loss of energy—to become bound together a little more strongly— and Jupiter imperceptibly increased its orbital velocity around the Sun.

Pioneer 10 was the first spacecraft on which a test of orbital acceleration was made, using a principle which might be called a gravitational sling or catapult (see Figure 59). For the space robots Voyager 1 and 2 this fuel-saving technique of gravitational steering is already practically a familiar routine. It can be used, however, not only to investigate the Solar System, but in principle for space travel too. Any spaceship, irrespective of its method of propulsion, can in theory profit by it and shorten its journey time. The idea goes back to the days of Sputnik, and came from Freeman J. Dyson.[9] The method is as follows. The spacecraft at first moves fairly slowly towards a binary, and approaches one of the stars in such a way that the craft and this star are moving precisely in opposite directions. Then the spacecraft swings around the star, and leaves the system with markedly increased velocity. In fact, the spacecraft departs more than twice as fast as the star it has rounded moves in its binary system, by taking energy from it. The two stars close up slightly and move a little faster. Subsequent travellers can profit in the same manner. But the whole process can only be repeated with one binary star system up to the point where the two stars come into contact, collide and destroy one another.

The following procedure can be thought of as an alternative. The spacecraft itself does not enter the binary star system, but sends a smaller body ahead on this path. When this body is catapulted out of the stellar system, it again encounters the mother ship, which has moved in the mean time to the anticipated rendezvous point. When the faster body is recaptured, its kinetic energy is transferred to the mother ship.

What speeds can be reached in this way? This naturally depends on the binary star. The distance between the two components must not be too small—so that the spaceship can pass between them without risk—and not too large, or the effect will be too small. Binaries with compact components, white dwarfs or neutron stars, are therefore the most suitable. As an example, two stars each of the mass of the Sun, with a separation of a few tens of thousands of kilometres, would orbit one another in a few minutes. A spaceship without propulsion of its own would be accelerated to 2000 kilometres a second by such a system; this is less than 1 per cent of the velocity of light, and corresponds to an effective acceleration of $10\,000g$! And the beauty of it is that the ship and its crew would survive quite uninjured: as a result of the gravitational field, the spacecraft is moving in free fall. The enormous effective acceleration by the gravitational field would thus go completely unnoticed. The only effect produced directly on the accelerated body by the gravitational field is the tidal force, caused by the decrease in strength of the gravitational field over the extent of the body (the 'gradient' of the field). For this reason the body cannot be too large, otherwise it will break up. A ship 100 metres across would only undergo a stress of $1g$; larger ships should not go around binary stars too closely. An interstellar space expedition could plan its route by the positions of close binary stars, so as to take in as many binaries as possible on the trip. Nor are smaller 'fishes' to be despised, such as stars with massive planets like the Sun–Jupiter 'binary system'. In Arthur C. Clarke's novel *Rendezvous with Rama* this possibility is put to use: the spaceship Rama, in its close flyby with the Sun, takes up not only solar energy but gravitational energy too.

It is unfortunate that the binaries which would be ideal for the purposes of space travel are very hard to find astronomically. White dwarfs are very faint. Neutron stars—and very probably also black holes—have already been found in some binary star systems, where they have been detected by their X rays. But since in these systems the stellar partner is usually of the 'blue supergiant' type, as in the object Hercules X-1, it is doubtful whether a spacecraft could safely navigate between the exotic pair of stars. The interstellar requirements are more probably fulfilled by the binary, which has as one component the radio pulsar PSR 1913 + 16, discovered in 1974. Although the other partner cannot be seen, it is probably at most as large as a white dwarf. Between them there is 'clean air'; there are no gas streams, and no hard X rays or gamma rays would disturb the process of 'filling up' with gravitational energy. Certainly the gravitational energy of binary stars can be obtained cheaply, but even in favourable conditions not more than 1 per cent of the velocity of light can be achieved by it. Besides this, the spaceship must not approach the binary too fast, otherwise it will not be sufficiently diverted from its path in the gravitational field of the stellar catapult, but will simply rush past it. So the space travellers must have plenty of time.

It is conceivable, however, that gravitationally assisted journeys between the stars might be undertaken with inhabited comets. As mentioned elsewhere (see Part II: 'The astroengineers of supercivilizations', page 200) Dyson thinks it possible that space colonies can be established at some time on asteroids and comets. The Solar System is surrounded by a spherical cloud of about 100 000 million (10^{11}) comets. This cloud, named after the astronomer Oort, is up to 2.5 light-years from the Sun. Although they have a total mass only a tenth of that of the Earth, the comets together have 1000 times the Earth's surface area. Comets are also rich in the important elements carbon, nitrogen, oxygen and hydrogen. The question is how common comets are in the Galaxy, and how many comets have been shot into the space between the stars, perhaps in the process of star formation. Of course, comets were already observed in ancient times, but they have been systematically studied only since the discovery of the telescope, and particularly since the introduction of photographic astronomy at the end of the nineteenth century. At present three or four comets a year are traced within the orbit of Jupiter. Comets arriving from interstellar space must have parabolic or hyperbolic orbits which reach to infinity. (Elliptical orbits are closed, and thus bound to the Sun.) But hyperbolic orbits can also be produced from orbits which were originally elliptical, by the influence of Jupiter and other planets. Such orbital changes have been observed for many comets. However, it is not out of the question that some few comets may have originated outside the Oort cloud. But these have not yet been specially studied for signs of life and the associated characteristics.

The black hole route to Andromeda?

Here I must throw cold water on an exaggerated hope which keeps returning to haunt books and minds, a hope engendered by the idea of space travel aided by black holes. The reason why black holes are so often responsible for all sorts of nonsensical speculations arises from the air of mystery given to these objects by scientific journalism, which rests on a simple lack of understanding. Even the astronomer Carl Sagan, a specialist in extraterrestrial subjects, falls victim to this error. First he says that ' . . . one conjecture . . . has been made, which cannot be disproved and which is worthy of note: black holes may be apertures to elsewhen. Were we to plunge down a black hole, we would re-emerge . . . in a different part of the universe and in another epoch of time. We do not know whether it is possible to get to this other place in the universe faster down a black hole than by the more usual route.' And Sagan concludes from this that black holes might be 'the transportation conduits of advanced technological civilizations—conceivably, conduits in time as well as in space'.[19]

Granted: it is an alluring prospect. Find yourself a black hole, disappear into it, and come out again into daylight or starlight beyond the Andromeda Galaxy. Of course it is conceivable, on a purely mathematical basis, that the Universe is put together in a more complex way than cosmologists normally suppose. This can be illustrated by the example of a plane surface. A flat plane is 'topologically' simple. But if we now construct a system of burrows in the plane, which go in at one hole and come out at another, we then have a topologically complex situation. Mathematically, similar tunnels can be constructed between two spatial points in our three-dimensional Universe. Relativity theorists call these features 'wormholes'. But black holes are unsuited to be both the entrances *and* the exits of such communicating passages. For example, if the entrance were a black hole, anything could indeed disappear into it, but nothing could come out again—unless the other end of the passage offered an inverted black hole, or a 'white hole'. Anything can *come out* of a white hole: but *what* will come out cannot be predicted physically on the basis of our current theoretical understanding. Even if there is a topological connection in the Universe with a black hole on one side and a white hole on the other, all information and every spacecraft would be destroyed in transit. And our physical experience so far provides absolutely no evidence of the existence of white holes. Astronomers believe that they have already observed black holes; whether white holes exist is still an open question. But the existence of white holes is generally doubted on account of their capacity for producing the unpredictable, with catastrophic results for physics. On top of this, nature has hitherto always shown herself to be topologically simple, and has given not a single hint that the Universe may be internally connected in some more complicated way (by additional 'burrows'). That it could ever be possible to plunge into a black hole, and then—in whatever condition—to return to this Universe, is contrary to theory and observation. It seems that if black holes have exits, they are not of this world.

Colonization of the Galaxy

A flight through the Universe at the headlong speed of 10 000 kilometres a second: of all the spaceship propulsion systems described in the last section, propulsion by hydrogen bombs, Project Orion, seems to be the most concrete. Certainly, it can already be made reality with today's technology. Though the spaceship Orion would be 1000 times faster than the chemical rockets of the present, such a spaceship would only cover in one century the short distance (by interstellar standards) of 3 light-years. Even at this rate, of course, it would be possible, eventually, to colonize the Galaxy, given a long lifetime for human culture. But how long would it take?

This was examined in 1976 with a computer experiment at the Research Center in Los Alamos; the results of the colonization study were published by Eric M. Jones. A civilization with space travel will have colonized the Galaxy in 5 million years.[20] Jones assumed that the mother civilization practised population control, and had an emigration rate comparable to conditions on Earth. Certainly the speed with which the wave of colonization spreads will not be greater than that of the spacecraft themselves—colonization at this maximum rate would already have embraced the entire Galaxy in 1 million years. Population control slows down the speed of expansion. At present the Earth's population is increasing—practically uncontrolled—by 3 per cent a year. During the settlement of North America between 1700 and 1790, less than one-twentieth of 1 per cent of the population of Europe emigrated. Jones used these assumptions as an analogy for space colonization: 'It [the occupation of North America] required a fairly sophisticated technology, and emigration was largely voluntary.'[20] Space colonies are expected to accept immigrants until they themselves are sending emigrants to new colonies which are still open. In this way, the colonies will continue to spread with about 15 to 25 per cent of the speed of the spacecraft, hardly slower. After 5 million years the Galaxy will have been crossed and colonized.

If the emigrants were to find fewer habitable planets than in Jones's calculation, this would simply increase the average rate of settlement, since the ships would then make fewer stops. Since the Earth itself was *not* colonized, Jones concluded, as Michael Hart had done before him,[21] that up to now no technological, spacefaring and colonizing civilization has arisen. In so far as this conclusion is correct, it provides additional support for the thesis that we are the only technological civilization in the Milky Way.

Here we should make quite clear that the population crisis *cannot* be solved by interstellar emigration. Even if we ignore for the moment all the technical problems and limitations, there is an *absolute upper* limit beyond which emigration to the stars can no longer keep pace with a population explosion. It can be shown that this follows from the physical fact that nothing moves faster than light; a wave of emigrants can thus spread at a maximum speed equal to that of light. An advanced culture with spaceships as fast as this, and growing in population by, for instance, 2 per cent a year, must—if it is to deposit its increase in space colonies—expand its settlement every year to the planets within a 2 per cent larger sphere. This possibility is exhausted as soon as the sphere must expand with a velocity greater than that of light in order to absorb the exponentially growing population surplus. It may be surprising, but these limits are already reached when the sphere is 300 light-years in diameter. Assuming that in this volume there are 30 000 habitable planets, an increase in population density is inevitable within 500 years. Even at the speed of light there is no escape from exponential growth: 'Interstellar expansion cannot solve the population problem, even with perfect technology; it only postpones it for 500 years,' comments Sebastian von Hoerner in the Weizsäcker Festschrift.[22]

In this discussion of the possibility and feasibility of space travel and the colonization of the (near or distant) Universe, I consider one more point worthy of notice. I should like to reduce it to the short formula: 'Space travel: now or never.'

I believe that civilizations like mankind reach a short phase once in the history of their development, in which it is practically decided once and for all whether space travel will ever be practised. Mankind is now in this phase. It is not so much a question of the exact time when a civilization will reach this phase, whether now or—if, for instance, we had not had the aid of Greek culture—a few centuries later. What is important is that we must expect to decide now, or at least very soon, whether the planets and perhaps the nearer stars will ever be opened up to us. If after this present phase of technological development mankind suffers some kind of setback—whether from atomic war or increasing overpopulation—we would simply be unable to make another start on a space travel programme, except by making 'inhuman' sacrifices. Since the last century we have been consuming the raw materials (metals, coal, oil, and so on) accumulated by the Milky Way in the interstellar gas before and during the formation of the Solar System, and later assembled in the Earth for thousands of millions of years by the Sun. In so doing, mankind is allowing itself an unrepeatable luxury. In 100 years most of the important materials which are now being consumed without restraint will come to a quick end (we call this the energy crisis). Some rare metals will already be exhausted before the end of this century.

Thus, a second start on space travel, in perhaps 200 years, will no longer be possible. The fossil and metallic raw materials, scarce but still available, would be needed for much more important, purely Earthly, purposes than for such things as space stations and spaceships. It will be useless, too, at this stage, to point to the unused metal deposits on the Moon or the asteroids. There will then be no rockets capable of lifting a minimum of 20 000 tonnes of implements and a nuclear reactor to the Moon, in order to even think of exploiting its wealth.

Nor should we speculate about solving all our difficulties, in this future without terrestrial raw materials, with glorious energy sources which would present no problems. The history of technology teaches us that for every device by which we have extorted an advantage from Nature—from horse to aeroplane to atomic power station—mankind has had to pay the price of burdening the environment and the irreversible use of raw materials. Indeed, the energy needed for each new and more complex means of transport and each new type of power station has shown a steady and rapid increase.

But if we are to see any advantage, even if only in the long term, in a programme of space travel, perhaps by the exploitation of ores on the Moon and the asteroids, it is only within the next few decades that we have a chance to build the superstructure which is indispensable for a venture into space, or at least into the relatively nearby Solar System. Space travel today is an ostentatious luxury. Society may well find good reason for *not* indulging in space travel on the grand scale. To cancel it would certainly be no great loss to civilization, if for instance it seemed likely that even in the long run mankind would have to invest far more than it could ever recover. Although cancelling the space programme might be to the advantage of coming generations, however, the chance of space travel would then be irrevocably lost. This also applies to Gerald O'Neill's project for building large space stations within the two-body system of Earth and Moon.

Space colonies in Lagrangia

We may see it as early as 1988: the first 10 000 settlers may by then have moved into Space Colony Mark 1, 300 000 kilometres from the Earth, near the Moon— a 500 000-tonne structure of aluminium, glass, titanium, soil and rocks, machinery, generators and other equipment (Table 14). The structure would be roughly cylindrical, but with many bulges, outworks, shafts, airlocks and solar mirrors. It would be 1 kilometre in length, 200 metres in radius, and would turn about its long axis every 21 seconds. On its outer wall each colonist would then have his normal Earth weight. And, in the year 2008—if it went according to Gerald O'Neill, the most enthusiastic propagandist for space colonies—10 million people could already be expatriated in space between Earth and Moon, almost completely self-sufficient in an environment which might offer them more comforts than their previous life on the over-populated Earth.[23,24]

The line of march is clear. In a construction phase a work station would first be erected, whose main component would be a sphere 125 metres in diameter. At nearly two revolutions per minute, the equator of the sphere will have half the Earth's gravity. For work under weightless conditions, 'it seems that . . . a daily interval of several hours at a gravitational level about a third of that of the Earth is necessary to prevent the loss of bone calcium'.[24] Large mirrors reflect sunlight through the windows of the sphere, the climate within the sphere being regulated by adjustment of the mirrors. Solar energy is much more effective in space than on Earth, where it reaches us only in a weakened form. A mirror in space weighing 1 tonne and 100 metres in circumference can provide power worth a million US dollars every year.

O'Neill also offers a plausible solution for the provision of materials. For the most part, these need not be brought from the Earth, which would be a very expensive operation. Exactly the right materials are lying around, unused, on the Moon. Here O'Neill draws on the results of the Apollo missions. The analysis of

Table 14 Materials for Mark 1 model*

	Material from Earth (tonnes)	Material from Moon (tonnes)
Aluminium (containers, structure)	—	20 000
Glass	—	10 000
Water	—	50 000
Power station (100 megawatts)	1 000	—
Initial structures	1 000	—
Special equipment, machines	1 800	—
Soil, rock	—	40 000
Liquid hydrogen	5 400	—
Construction force (2000 settlers)	200	—
Dried food	600	
TOTAL	10 000	~500 000

* After O'Neill.[23, 24]

the Moon probes showed that the surface of the Moon is rich in iron, aluminium, titanium and magnesium—in fact, nearly 30 per cent of it consists of metals vital to the construction of energy satellites and space stations. Of the remaining lunar rocks 40 per cent consist of oxygen and 20 per cent of silicon: the atmosphere of the colony will come from this oxygen. The exploitation and transportation of lunar material is 20 times cheaper than from Earth, as the Moon has no bothersome atmosphere and a far smaller surface gravity. Conveyance into space is correspondingly easy. The material, slightly compressed, will be loaded on to a kind of sled and catapulted into space, magnetically accelerated along aluminium rails 10 kilometres in length, its path accurately corrected by laser radiation. The sled will be braked and return to the mining site in 150 seconds. The energy required for the sled traffic will be generated by an atomic power station erected on the Moon for this purpose, as the Moon's day, 2 weeks in length, is followed by an equally long night, cold as interstellar space. But this seems to be the only purpose for which nuclear energy will be necessary.

Calculations show that a sled catapult, far superior to any rocket, can transport up to a million tonnes of lunar material a year into space. Later it will also be possible to transport whole asteroids from the asteroid belt (between Mars and Jupiter) or to work them on the spot. It may seem surprising, but the energy required for transport from the asteroid belt to the space colony is no greater than if the material were carried up from the Earth.

Where is the space colony to be built? Geostationary orbits, in which the colony is 'suspended' at a distance of 36 000 kilometres above the same point on the Earth, have disadvantages. The orbit is in the middle of the Van Allen radiation belt, and is hard to reach from the Moon. There is however, another possibility. In 1772, the French mathematician J. L. Lagrange investigated the so-called 'restricted three-body problem'. This examines how three gravitationally interacting bodies will move. The 'restriction' consists in assuming that one of the three bodies is of negligible size. Lagrange showed that the small body can remain at definite equilibrium points in the gravitational field of the two other large bodies. In the Earth–Moon system there are exactly five of these so-called Lagrangian points. The physical explanation is as follows. As the Earth and the Moon orbit one another, the gravitational fields of Earth and Moon are complemented by the centrifugal force of the system. At the Lagrangian points, gravity and centrifugal force balance one another; they are thus suspended. The first (L_1) is between the two larger bodies, the second and third (L_2 and L_3) are outside on the line joining them, and the two others (L_4 and L_5) each form a more or less equilateral triangle with Earth and Moon and lie on the Moon's orbit, 384 000 kilometres from both Earth and Moon. These last two Lagrangian points have the advantage in the Earth–Moon system of being stable. This means that a small disturbance to the orbit would not force a body completely away from the point, but would drive it back to it or cause it to take up a small orbit around the point. In O'Neill's first plan, therefore, L_4 and L_5 were chosen as the sites for a colony. (In the United States a Society concerned with the L_5 colony has even been formed.) Orbital perturbations by the Sun will still prevent a precisely stationary positioning of a colony at L_4 or L_5. Instead it would move in a slightly kidney-shaped path about L_4 or L_5, 140 000 kilometres away (see Figure 60, top).

Calculations by T. A. Heppenheimer, however, showed that L_5, though it looked good at first glance, was by no means the best choice.[17] The reason for this is connected with the transport of Moon material to L_5. For navigational reasons it is preferable, and therefore envisaged, not to carry the material to L_5 directly, but on a circuitous route through L_2, 60 000 kilometres beyond the Moon.

The detour through L_2 has an advantage compared with a direct flight to the space colony. If, when the lunar material is catapulted, in spite of accurate adjustments, there are variations in the velocity of ejection (harder to correct than the direction of firing), the lunar packages will nevertheless still reach L_2. For direct flight, complicated course adjustments must be effected by means of auxiliary rocket steering. But in the neighbourhood of L_2 the gravitation of Earth and Moon acts like a lens, which still directs the craft with small deviations of velocity towards the 'focus' L_2. Incidentally, the best firing site from which to reach L_2 from the Moon in this way lies near the craters Censorinus and Maskelyne, at 33.1° East longitude.

According to the plan, a 'mass trap' will wait at Lagrangian point L_2 to

Equilibrium points in the Earth-Moon system

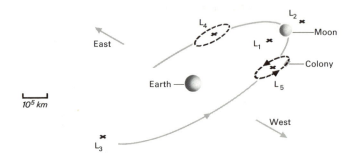

Resonance path with transfer orbit from L_2

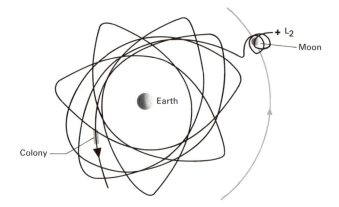

Figure 60 Top: Location of the equilibrium points (Lagrangian points) in the Earth–Moon system. Not all of these are stable; for example point L_1 is unstable: small perturbations will cause a body momentarily at rest here to fall either to the Moon or to the Earth. However, L_4 and L_5 are stable, and space colonies could be placed there. But constant perturbations by Sun and Earth would result in the colonies taking up a nearly elliptical path about these points (broken lines). Bottom: An alternative orbit for a space colony is offered by the '2:1 resonance orbit' of the Earth–Moon system. Its advantage over L_5 is that considerably less energy is needed to transport material via L_2 from the Moon, or even from Earth, to this orbital path. (After ref. 17.)

Table 15 Possible stages in construction of a space colony*

Model	Length (kilometres)	Radius (metres)	Period (seconds)	Population	Earliest completion
Mark 1	1	100	21	10 000	1988
Mark 2	3.2	320	36	150 000	1996
Mark 3	10	1 000	63	1 million	2002
Mark 4	32	3 200	114	10 million	2008

* After O'Neill.[23, 24]

capture the 9-kilogram packet in a sort of net, 2 days after its firing from the catapult, when it is still travelling at 900 kilometres per hour. It will then take care of onward transport to L_5. For this the mass trap requires a change of velocity of 1700 kilometres an hour. This raised the question: might there be other equally stable orbits between Earth and Moon that the mass transporter could reach from L_2 with a smaller expenditure of energy and a lower velocity?

This question was studied by computer simulation in 1976. A body was released (mathematically) from L_2 and its path studied. The resulting orbit calculated by the computer was very interesting. After circling the Moon twice the craft took up an almost stable, somewhat elliptical orbit about the Earth, with an orbital period of just 2 weeks. A further finding was that a colony in this orbit, known as a '2 : 1 resonance orbit' (see Figure 60, bottom), could be reached from L_2 with a change in velocity of only 36 kilometres an hour! The former organizers of L_5 concluded: 'The colony will arrive at an orbit about the Earth, 360 000 kilometres away at the farthest, 160 000 kilometres at the nearest point.' Of course this choice of site does not rule out the colonization of L_4 and L_5; but these are more expensive, and therefore the second and third choices.

But what will the colony look like, and what will daily life be like in space? At present there are at least three major designs for space colonies (see Table 15). Besides O'Neill's cylinders, the 'wheel with spokes'—Mark 2—seems to have found most favour (Figure 61). It was arrived at during a Workshop on Space Colonization which took place at Stanford in the summer of 1975, where 28 scientists, engineers, sociologists and agricultural technicians had gathered for a 'brainstorming' session about O'Neill's space colony. Here the wheel concept, since designated the 'Stanford Torus', was born.[25] Life in the Stanford Torus, as in O'Neill's sphere and cylinder, will be very like that on Earth. The length of the day will be adjusted by the opening and shutting of the principal mirrors rotating with the colony. Even the seasons can be simulated by varying the lengths of the days. Birds and animals can be accommodated, along with lakes, small rivers and trees. Plants, since they would have grown in terrestrial gravitational conditions,

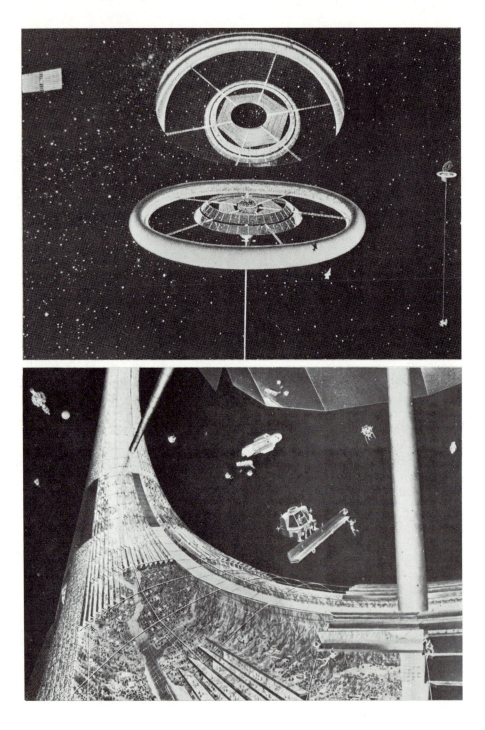

Figure 61 The second (Mark 2) type of space colony proposed by O'Neill resembles a wheel with spokes. Above this 'Stanford Torus' is an enormous mirror, enabling the colony to regulate its supply of solar energy. The lower picture gives a glimpse into the interior of the 'ringworld'. (Courtesy: NASA.)

would have the same forms as on Earth. In the Mark 4 version (see Figure 62) the weather would be summery, with 50 per cent humidity, a cloud limit between 1100 and 1400 metres above the wall of the cylinder (towards its axis), with temperatures between 32 °C and 0 °C. The agricultural areas, kept separate from the residential quarters, would be maintained under special conditions (illumination by day and night, controlled temperature, carbon dioxide content and humidity). Chickens, goats and cattle would be bred in farms. With adequate food and electrical energy, and a controllable climate with no smog—no combustion motors permitted!—life would be more pleasant than in many places on Earth. Houses will be built by a kind of honeycomb system. Within each unit of the colony, bicycles and slow electric cars would be sufficient transport. For communication, everyone will have the use of a computer connected to a central data network. There will also be scope for sports with far more possibilities than on Earth. But yet the colony will be no paradise. 'It will be comfortable and attractive, but it will still be a frontier community working hard to build power satellites,' says Heppenheimer.[17]

Safety seems to present no problems. Protective shielding will be built around the settlement against the dangers of cosmic radiation. The danger from meteors is small: most meteors come from comets, not from asteroids, and thus consist principally of frozen gas and dust. As judged from the frequency of meteor impacts registered on the Moon by the Apollo seismometers, and from the NASA observations, a space colony with a surface of 1000 square kilometres would be struck by a meteor weighing 1 tonne only once in a million years on the average. And even in that case the impinging body would inflict only local damage if the colony is well constructed. One impact by a meteorite of 100 grams is expected on the average every 3 years. On account of the greater frequency of grains weighing less than 10 grams, the construction of the windows must take particular account of impacts of this kind. But O'Neill calculates that leaks need cause no panic: it would take a relatively long time for the loss of air to become noticeable. 'Provision against meteor damage is more a matter of sensors and regular small repairs than of sudden emergencies,' says O'Neill.[24]

Hitherto there have been few critical remarks, like those of A. V. Cleaver, about the colonizing programme.[25] And even the critics do not deny the *technical* feasibility with present-day capabilities. Doubts have principally been raised as to the calculated cost, and the point in time at which the space colony can realistically be expected to be circling the Earth in the 2:1 orbit. Here indeed mankind has a project in its repository of brilliant ideas which seems to be

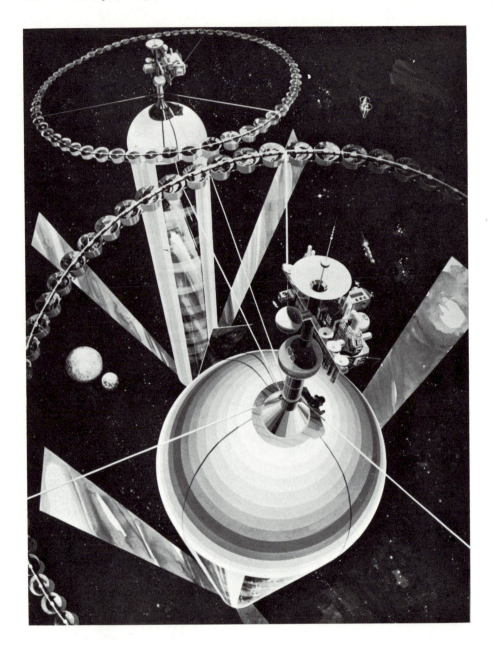

Figure 62 If Princeton physicist Gerald O'Neill's plans come true, several million human beings could one day live in this space colony, shaped like a double cylinder. Each cylinder of the Mark 4 model would be 32 kilometres long and 6.4 kilometres wide, and would turn on its axis once every 2 minutes. (Courtesy: NASA.)

feasible, but about which the question arises whether, when and why it should actually be put into effect—for even the largest space colony cannot contain the population explosion. For this we must find a solution for ourselves here on the Earth.

A more important aspect than the colonization of the reaches of space between Earth and Moon may therefore at present be the establishment of industrial manufacturing processes in space conditions. Exotic structural materials such as steel foam can only be produced in weightless conditions, in the free fall of a space station. In addition, the provision of services is possible, such as the construction of satellite power stations, a project which has been pursued since 1968 by the American engineer Peter Glaser. This involves the capturing of solar energy with mirrors, its conversion into microwaves with an electric generator powered by helium gas, and its transmission to Earth. The receiving station, an arsenal of aerials several kilometres across, would convert the incoming rays back into electricity. In astronomical research, too, there is a need for observations of the sky from stations outside the atmosphere. The information represented by the Skylab pictures is already of great value. So there is already a need for spin-off benefits from space travel in the fields of industry, power generation, scientific research, perhaps even for medical purposes. For these at least there is already definite motivation.

Gerald O'Neill is said to have hit on his first idea while lecturing to freshmen at Princeton in 1969. There he raised the question: 'Are the surfaces of the planets the right place for an expanding technological civilization?' I think that (at present) the answer must be 'yes'.

UFOs and anti-gravity

They fly faster than an aircraft could ever do, hang suddenly motionless in the air, and then shoot away. They emit a strange hum, approach silently, sometimes even fly in formation. They appear to the astonished eye as elongated or saucer-shaped objects. They eject gases or move apparently without propulsion, are accompanied by luminous phenomena and are detected on radar screens. It is not only professional 'ufologists' who report these astonishing phenomena, which are today classed under the collective name of *U*nidentified *F*lying *O*bjects. Sober citizens, even hard-boiled test pilots, radar technicians and scientists have also reported sighting something they cannot explain—a UFO. Not only that: in some cases they have waved at the creatures who emerge from these strange craft, and even enjoyed a short ride in the extraterrestrials' spaceship. (Indeed, a Brazilian farmer—at least so he says—was dragged into a ship and *forced* to sleep with a bewitching UFO woman.) And the world need not rely solely on the verbal testimony of the witnesses (or victims?). They often offer photographs and films, or point to the landing places—scorched earth—of the spaceships.

The final sections of this book will not, therefore, be dedicated only to those who have known it all along—that we have long had contact with extraterrestrial civilizations and have been visited and observed by them, or that at least in the dim biblical past they have descended on the Earth—but also to those among our sceptical contemporaries who, from their own experience or through reports they were driven to believe, have become convinced that UFO phenomena represent a genuine visitation by extraterrestial beings. We shall not dismiss this subject superficially; we shall not therefore discuss the more than 99 per cent of all UFO reports which can be fairly quickly explained in terms of 'natural' phenomena: aeroplanes of unusual type or seen in exceptional weather conditions (clouds and luminescence behind the machine) or with atypical lights; weather balloons, satellites, planets—especially Venus—or bright stars; flocks of birds, luminous organisms, headlamps reflected from clouds, sunlight reflected from shining surfaces; ball lightning or meteors.[26,27] In the United States alone more than 10 000 UFO cases have been recorded since 1947, when the UFO wave got under way. The American magazine *National Enquirer* even picks out the 'UFO Case of the Year' for a prize.

For the sceptic in UFO matters there is only one place to start: it must be shown beyond doubt that at least *in a single case*, extraterrestrials had a hand in the game. The evidence for the UFO hypothesis can *not* be produced by an accumulation of more or less mysterious phenomena or archaeological finds. Quantity does not confer conviction. Many 'maybe's' can never add up to one 'definitely'; the contrary is true. A scientifically acceptable proof need only be demonstrated in one case. Therefore I shall only describe here *one* chosen case as representative of many.

UFO attacks US copter

The case we shall be dealing with here was chosen, in 1973, in a US$5000– competition as 'Best UFO Case of 1973'; it won the prize, and was at the time declared 'inexplicable' even by critical experts. To anticipate the result: it can be reduced to an earthly cause, albeit after much detective work and expenditure of time—as incidentally can all other 'famous' cases. 'The Best UFO Case of 1973' took place on the night of 18 October. 'The four-man crew of a U.S. Army helicopter flying in the neighbourhood of Mansfield, Ohio, . . . had a terrifying encounter with a brightly glowing object moving at great speed and risking a mid-air collision [with the helicopter].' Thus begins the description of the incident by the scientific author Philip J. Klass in his book *UFO's Explained*, in which he critically examines the best-known UFO events.[28] In the following I largely follow his analysis.

First let us have the story. The incident occurred at the high point of a wave of alleged UFO sightings in Ohio. The helicopter was flying home from Columbus to Cleveland, in clear weather, at an altitude of 800 metres above sea-level and at a speed of 160 kilometres an hour in a north-easterly direction. A few minutes after 11 p.m. the pilot saw a bright red light to the east, which he took at first for the red warning lamp of a television tower. But then the light seemed to be moving parallel with the helicopter. It grew larger and brighter, and approached on an apparent collision course. Says the pilot Lawrence Coyne: 'I looked through my right-hand window, and saw the light approaching at high speed, more than 1100 kilometres an hour.' Coyne tried to radio the nearby airport at Mansfield to find out whether perhaps an F-100 jet fighter was just landing, but could not get through to the control tower. To avoid a crash Coyne flew lower, at first falling by 300 metres a minute, and then, when the object continued heading for the helicopter, with a speed of descent of 650 metres a minute, in a 25-degree dive. But despite all these manoeuvres the shining object still seemed to be coming straight for the helicopter: the crew prepared for a collision. But there was no collision. Instead, the puzzled soldiers saw the object above the helicopter, where it seemed to be stationary. Coyne had the impression that it was '15 to 20 metres in size . . . the front end bright red. The rear end had a green light, and you could see where the light stopped and the grey metal structure began . . . The light at the end of the ship turned through 90 degrees and pointed down at the helicopter. . . . and flooded the cockpit with green light.' Coyne estimated that for a few seconds the object hung as low as 170 metres above the helicopter. However, they heard no sound of a motor or anything of the sort. Then Coyne saw from a glance at the instrument panel that they were suddenly flying at a height of more than 1100 metres and were rising at 300 metres a minute. Although they thought they were still descending, they had actually risen without knowing it while they were observing the object. They made vain attempts to establish radio contact with three nearby airports: 'We could neither transmit or receive!' Finally, after 6 minutes they received signals from a control tower. They landed in Cleveland, and there they told of their experience.

Now for an analysis of the events. An F-100 jet was not involved. The last one had landed at Mansfield air base at 10.47 p.m., a quarter of an hour before the event. In November 1973 Klass, whose attention had been attracted by a talk show on which Coyne appeared, investigated whether it could have been a glowing meteor—though this, of course, would not have explained the rapid and unnoticed rise of the helicopter, nor the brief failure of the helicopter's radio. Yet there were parallels with other meteor sightings, when experienced pilots at night had underestimated the distance to supposedly nearby meteors by as much as 200 kilometres. There too the pilots had undertaken manoeuvres to avoid the supposed danger of collision. Experience shows that the distance of unusual objects is almost always misjudged, especially at night when there is no reliable guide. Meteors, especially those which enter the atmosphere almost horizontally, are often taken for UFOs—like the 'Iowa meteor', which flew over Illinois and Iowa on 5 June 1969 and burned out (Figure 63). A high-flying

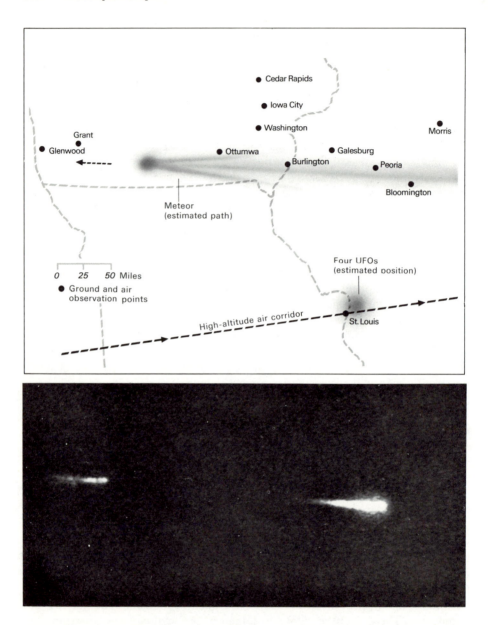

Figure 63 The 'Iowa Meteor' and its debris were observed simultaneously on 5 June 1969 by three experienced flight crews in the air near St Louis. The crew of an airliner took the luminous phenomenon for a squadron of UFOs on a collision course. After the actual path of the meteor had been reconstructed—it flew almost horizontally for about 500 kilometres over Illinois and Iowa—it was found that the pilot's estimates were out by 190(!) kilometres. The photograph shows the meteor as it flew over Peoria with its fragments; the map indicates the error in the estimate of the plane crew. (After ref. 28; photo: *Peoria Journal Star*, by Alan D. Harkrader, Jr.)

meteor would at least explain the absence of noise. Meteor showers are also typical of the second half of October, when the Earth crosses the path of the so-called Orionid meteors. A long, glowing meteor tail would illuminate the cockpit for several seconds, and could easily give the impression that a glowing object was *motionless* above the helicopter. As for the green light, the transparent roof glass of the helicopter was coloured green for protection against sunlight.

Now to the unnoticed ascent, Coyne later conceded that in a moment of panic he might instinctively and unconsciously have jerked the control lever upwards. On this point his recollection was wrong: certainly the helicopter was originally flying 600 metres above sea-level, but it was only 130 metres above the ground! It would have crashed to the ground in the first 12 seconds of the avoidance manoeuvre. Investigation of the rotor blades and of the helicopter, moreover, showed no signs of damage such as a rapid updraft would have caused. Coyne's statements of the time taken for the descent and ascent were also incorrect—a consequence of shock. Klass was also able to explain the failure of the radio equipment on board the helicopter. As a test, Coyne repeated the flight, and again attempted to establish contact with the three nearest air bases, at the same low altitude, from the position at which the 'UFO' had appeared. This time, too, he failed to get through. From there, they were in 'radio shadow'. On the first attempt to establish contact with the Mansfield air base, it probably had simply not replied; this happens all the time in flying. It is also possible that the radio operator changed channels too quickly, before the equipment could adjust itself to the new frequency. For the transmitting equipment in question this takes up to 5 seconds. Thus the case of 'UFO attacks helicopter' was resolved into a chain of at least very plausible occurrences. No single stage of the event compels us to bring in an extraterrestrial spaceship to explain it.

Klass was also able to explain cases that attracted attention in the 1950s, in which UFOs were seen on radar screens. These were chiefly cases of peculiar reflections of radar beams. Klass's investigations show that, as he himself writes, 'many UFO cases seem mysterious and inexplicable because the people who investigate the case put too little solid work into the investigation'. This is hardly surprising. Most ufologists 'prove' what they themselves, consciously or unconsciously, wish to believe. The German engineer Adolf Schneider is one of these, with his book *Besucher aus dem All—Erforschung und Erklärung des UFO-Phänomens (Visitors from Space—Investigation and Explanation of the UFO Phenomenon).*[29] He is mentioned here only as a representative example.

Schneider was particularly concerned about gravitation (or rather anti-gravitation). Appealing to an obscure 'theory' put forward by a Japanese, he adopts the principle, totally incomprehensible in his text, of an 'anti-gravity drive' with which an 'artificial gravitational field' can be produced, by which 'a transformation from gravitational potential energy into mechanical or electrical energy takes place'. Schneider believes that many UFO phenomena confirm the use of this 'drive'. On page 312 of the 'third, revised edition, 1976' are the thought-provoking words: 'Particularly impressive evidence for the operation of an anti-gravitational field is offered by the report of the crew of an American helicopter.' And he comments: 'There are reliable reports of this kind from many countries. They impressively confirm the theory that there must be spacecraft which produce a uniform anti-gravity field and are even able to extend it for some distance.' This refers to the case, discussed above, of the helicopter crew in distress—here described from the standpoint of a True Believer!

Another case was chosen from among 1000 entries to a competition in 1973 by a jury including the astronomer and author J. Allen Hynek,[30] as 'a great scientific mystery—the most puzzling case that the jury has encountered in a full year of investigation'.[28] The family of a Kansas farmer claimed to have seen a 'mushroom-shaped' object near the farm in 1971. It hovered over the ground between some trees for some time belching flames, and then flew off with a screeching motor. A year later, Klass examined the alleged solid evidence for the experience, a ring-shaped patch of whitish earth at the spot where the UFO had paused. Ruling out the consequences of a fire—there were undamaged branches on the ring of earth the day after the occurrence—he was able to show that very probably a circular food or water trough must have stood there, perhaps for the sheep on the farm. A Polaroid picture of the spot, on which the farmer had recorded the alleged afterglow of the ring in the darkness, was shown to be a fake: it had been taken with a pocket lamp under the full moon. The jury was taken in by the fraud of a farmer who found himself in need of money.

Psychologists have shown by special experiments how easily UFO occurrences could be fabricated, which people will nonetheless 'buy', country-wide and world-wide. And with a little skill, good photomontages of UFOs can be made. Besides this, we should take note of the fact that hitherto no reputable observatory has ever recorded a flying object which has had to be permanently recorded as a UFO.

But what about anti-gravity? For there is an important difference between electric fields and gravitational fields. An electric field has two types of charge, positive and negative: similar electric charges can enhance one another and opposite electric charges can cancel one another out. But gravitation has only one type of 'charge': the mass of the body. (Anti-matter has the same gravitational properties as matter.) Therefore masses can only mutually increase their attractive force—as similar electric charges do—but never cancel one another. This is why the Sun has a greater gravitational field than the Earth. To neutralize the gravitational field of a body would need a kind of matter with 'negative mass', which has never been found up to now. But to return to the UFOs . . .

UFOs, thinks the plasma physicist P. A. Sturrock of Stanford University, will probably be recognized eventually as:

1. A fringe phenomenon, composed of tricks and hoaxes using known phenomena and objects. Or

2. A real phenomenon, consisting of unusual manifestations of known objects and events. Or

3. A real phenomenon, which has up to now been neither recognized as such nor understood in the framework of accepted science.[31]

Each of these possible endings to the story holds promise of insight. If we eliminate hoaxes and misinterpretations, everything speaks in favour of (2). Case (3), the UFO hypothesis, should only be kept in mind until the time when it can be shown, beyond the possibility of doubt, that the landing of an extraterrestrial spaceship has actually taken place on Earth. Of course this would still not prove that UFO experiences hitherto have related to actual occurrences in the sense of the UFO hypothesis.

Until that time it seems that the solution of the UFO problem lies more in the area of mass psychology, especially at times of anxiety and political insecurity. Perhaps it is no accident that UFO sightings accumulated most feverishly in the United States in the 1950s, during the Cold War, and again the late 1960s, during the Vietnam War.

UFOs: the modern myth

Mankind has always had a defence ready against apparently overwhelming threats: mythology. The danger is at once lessened if—on whatever basis—it can be 'explained'.

Myths were the earliest 'explanations' of many of the potential threats to life in the world, long before the coming of science. The more strange facts could be embraced by it, the more powerful was the myth. The creation of myths is deeply rooted in the hearts of men. Myths are still being created today, relatively untouched by scientific theories. Some of the older myths, however, have vanished, and been resolved into scientific explanations—not always to popular satisfaction. The myth of the thunder god, who hurls lightning and hail in fury on the Earth, may be psychologically more acceptable to some people than strange stories about electric discharges and supersonic shock waves spreading through the atmosphere. The same may have happened with the UFOs. For there was an old, old myth that once, in the dawn of world history, the gods came down to Earth and left the seeds of life behind. Now, with our knowledge of air transport and genetics, this myth—after the first wave of fear over UFOs in 1897—has been dug up again (in 1947), and rejuvenated with small changes. Not only did the gods visit us—they are still among us! They are observing us, if not indeed watching over us. But clearly they wish to remain anonymous in their activities, though some people actually claim to have had direct contact with them. This is the UFO myth. It has so strong an influence on people's minds that most of them think first of a UFO—irrespective of what sort of phenomenon they witness—

before they hit on explanations in terms of things that are usual today, such as weather balloons, satellites, reflections of the light or meteors. The fact that the enormous majority of UFOs are very quickly explained in quite Earthly terms shows how ready many people are to accept the myth rather than any scientific explanation.

Why is this so? Many of the scientific explanations offerred for these frequently very impressive UFO events are often quite complicated or trivial. They bore people, they give no scope for fantasy, and can in no way compete with the romantic UFO dogma! At the price of stimulated emotions and science-fictionesque pseudoreligious ideas, there are many people who will readily renounce scientific argument or a strict proof (or disproof). And how superior the UFO myth shows itself to be from a psychological viewpoint: how many phenomena it can be applied to at one go! It can be invoked to explain countless everyday phenomena so long as no immediate counter-explanation is available. A conventional explanation, on the other hand—as Philip J. Klass has proved—can often only be achieved after much hard detective work. The first major study, the $500 000 'Blue Report' by the Nobel prizewinner Edward U. Condon (1968), had to leave the question open in 34 cases.[26] Even those who are scientifically inclined or trained feel that they are entitled to apply their (mythological) beliefs to these 'gaps'.

Not that scientists can stop them. Indeed in such cases they have (at first) nothing better to offer, unless someone like Klass, for example, is to pursue further months-long investigations. They can only advise caution in rushing to draw conclusions which for various reasons are rather improbable. In the continuing absence of solid positive evidence, the UFO hypothesis is no better than the ancient idea of the noisy thunder god.

As a philosophical tool for cases which sow such confusion, scientists make use of the principle of 'Occam's Razor'. Occam, a scholastic theologian of the thirteenth century, who came into conflict with the English church and found political asylum in Munich, was an opponent of metaphysics, and made important contributions to logic. Occam's Razor gives a criterion by which *one* explanation can be systematically selected when several are offered at the same time. The criterion always favours as the most probable the explanation which rests on the smallest number of assumptions. Although the grounds for applying this selection principle are ultimately of an aesthetic nature, it must be said it has hitherto worked remarkably well in science.

When applied to the UFO problem, Occam's Razor would counsel against drawing any hasty conclusions for new and unexplained cases. Until someone provides proofs, the UFO hypothesis can be kept in hand.

The hypotheses of Erich von Däniken must be looked at from a similar point of view. 'I maintain', he says, 'that this planet Earth received visitors from space more than once in the early days of history. These beings from alien galaxies changed our forefathers by operating on their genes, otherwise we should still be living among apes today. They created man in their image. These visits are echoed in folk legend and myth.' These words of Däniken, spoken in a lecture, are quoted by Günther Rühle in the *Frankfurter Allgemeine Zeitung* of 6 April 1977.

Däniken finds 'confirmation' for his speculations above all else in certain archaeological finds, particularly in those which give evidence of astronomical knowledge which many find surprising, notably that of the Mayas, the Aztecs, the Incas, and also the African tribe of the Dogon. For instance, in Maya temples on the Mexican peninsula of Yucatán, it has been found that the foundations of one temple are aligned to the extreme points of the rising and setting of the planet Venus. This circular building at Chichén Itzá may once have served as an observatory (Figures 64 and 65). As the German engineer Horst Hartung and his associates found, the lines of sight from certain windows, diagonally from the inner to the outer corners, point in the direction in which Venus sets at its most northerly and southerly.[32] The precision is remarkable and implies that the Maya must have known the (synodic) cycle of Venus with an accuracy of 6 minutes. But this accuracy, easily attained with the naked eye, 'only' shows that the Maya, like other ancient civilizations, had already been making systematic observations of the heavens for centuries—an admittedly astonishing, but not impossible, cultural achievement. At all events, it did not require the promptings of extraterrestrial astronauts.

The Dogon and the mystery of Sirius

It is the same with the Dogon. How do the Dogon, a 'primitive' tribe in Central Africa, know that the Dog Star Sirius A is actually orbited with a period of 50 years by a companion, which is (today) totally invisible to the naked eye? The companion star, called Sirius B, was only found by astronomers in the nineteenth century, and was classified in the present century as a so-called white dwarf. This knowledge, which reason suggests should be quite unavailable to the Dogon, along with more strange astronomical information besides, was learned by the French ethnologists Marcel Griaule and Germaine Dieterlen in the 1940s, when they undertook a study of the Dogon south of Timbuktu. The ethnologists were amazed, and especially so when it became clear that this astronomical observation had been a tribal tradition for several thousand years. The young American orientalist Robert K. G. Temple took up this theme, encouraged by the science fiction writer Arthur C. Clarke, in his book, *The Sirius Mystery*.[33] He tried to get to the bottom of the mystery in the mythologies of Egypt, Sumeria, Greece and elsewhere. His thesis is that around 3500 BC the inhabitants of the Sirius system visited the Mediterranean region. In the course of time, tribes

Figure 64 Ruins of the Caracol in Chichén Itzá, probably an astronomical observatory of the Maya, seen from the south. (Courtesy: *Umschau in Wissenschaft und Technik*.)

driven into the region of the Dogon (perhaps the Garamantians) brought this tradition with them, and so the information found its way into the myths of the Dogon. But this speculation is untenable.

Temple's semantic and mythological investigations, going back to an Egyptian origin for the Dogon legends, seem less convincing than he thinks. The period of revolution of Sirius A and Sirius B sometimes turns up as 50 years, sometimes as 60. Moreover it is uncertain whether Sirius B is what is really intended by the Dogon's representation of Sirius as an ellipse with nine accompanying symbols.

The work of the astrophysicist D. Lauterborn, now at the Observatory in Hamburg, shows that knowledge of Sirius B could also have come to the Dogon in a simpler way. He tells us that at the time when Egypt was great, large masses of gas were still flowing between the two components of Sirius, which (as confirmed in historical documents) rendered the system clearly visible as a red star (Sirius A is still the brightest star in the sky). It is practically out of the question that the double star Sirius can have planets at all, let alone planets on which a spacefaring civilization might have arisen. Nevertheless it remains a remarkable astronomical achievement by the Dogon (or the Garamantians) to

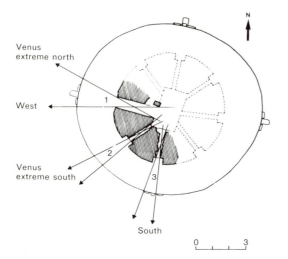

West

Venus
extreme north

Venus
extreme south

South

N

1

2

3

0   3

Figure 65 Plan of the remains of the observation chamber in the upper part of the Caracol. Sight lines and the new findings have been indicated in the extant windows. (Courtesy: *Umschau in Wissenschaft und Technik*.)

detect the change in the red star and its period of about 50 years with the naked eye.

But independently of how we regard such attempts at interpretation, the pleasing thing about Temple's book is that he makes no religiously tinged attempt to bulldoze his ideas through with 'proofs' based on protestations and repetitions, but presents his results in a spirited but unprejudiced manner. He does offer speculations, but they are clearly recognizable as such, and are kept distinct from the true research. Nowhere is the reader required to swallow them as dogma. Temple leaves him with the pleasant freedom to enjoy the interesting ethnological material, without having to accept his speculations about visits from 'extraterrestrials'.

Although unqualified from a scientific point of view, Däniken's propaganda campaigns should be looked at in the right light. From a purely literary point of view he can be regarded—along with the less successful originator of his ideas, Robert Charroux—as the founder of a new genre of fantasy literature, which one could perhaps call historical or documentary science fiction. Däniken has at all events attracted a following. Richard Kaufman, in the *Süddeutsche Zeitung*, has even dubbed it a 'new type of escapist literature'. It must also be admitted that in his enthusiasm for archaeological investigation Däniken has drawn attention to some no less shaky and dubious attempts at explanation by professional archaeologists, and to gaps and contradictions in their theories—which are otherwise perhaps quite acceptable. From another point of view, Däniken has founded a contemporary, up-to-the-minute religion. He is a very successful author, and his books are said to have attracted over a hundred million readers spread over all five continents, with 41 million books sold. As the critic Knut Herbert once put it, he is 'the founder of a profession which may one day be called "mythological engineering" '.[34] Däniken's writings are dangerous precisely because of their success, particularly at a time in which traditional religions are neglecting more and more to draw contemporary mythological ideas to themselves. The modern side of Däniken's polytheism lies in the fact that he demands no arbitrary act of faith from his apostles, but offers 'evidence'—as in the title of one of his books.[35] This dresses un-modern dogma in a white laboratory coat, and makes it scientifically and socially acceptable. It is for such reasons that Däniken's followers are willing to make a few logical leaps in matters of detail and get by without conclusiveness in his 'evidence'. For example he shows the runways of the first astronauts, driven to Earth after a galactic war, who then made their homes in the caves of Ecuador, and mutated apes into men. Here is his glib answer: 'They do not simply look like runways. They *are* runways.' Needless to say, here as elsewhere the proof is replaced by an assertion. But he who is at heart a True Believer will nonetheless call it science.

References

Part I (*pages 1–90*)

1. S. Weinberg, *The First Three Minutes* (Basic Books: New York, 1976; André Deutsch: London, 1977).
2. K. Pinkau, 'Die Entstehung der Elemente im Kosmos', *Die BASF* **1** (1977), 60.
3. B. Zuckerman, 'Interstellar molecules', *Nature* **268** (1977), 491.
4. E. Herbst and W. Klemperer, 'The formation of interstellar molecules', *Physics Today* **29**, No. 6 (1976), 32.
5. R. D. Brown, Interstellar molecules, galactochemistry and the origin of life', *International Science Reviews* **2** (1977), 124.
6. *Sterne und Weltraum* **11** (1977), 371.
7. R. A. Creswell, 'Moleküle im interstellaren Raum', *Umschau in Wissenschaft und Technik* **77**, No. 11 (1977), 361.
8. A. C. Cheung, D. M. Banks, C. H. Townes, D. D. Thornton and W. J. Welch, 'Detection of NH_3 molecules in the interstellar medium by their microwave emission', *Physical Review Letters* **21** (1968), 1701.
9. F. Hoyle and N. C. Wickramasinghe, 'Polysaccharides and infrared spectra of galactic sources', *Nature* **268** (1977), 610; *Lifecloud* (Dent: London, 1978).
10. V. I. Goldanskii, 'Interstellar grains as possible cold seeds of life', *Nature* **269** (1977), 583.
11. A. G. W. Cameron, 'The origin and evolution of the Solar System', *Scientific American* **233**, No. 3 (September 1975), 32.
12. D. D. Clayton, 'Solar System isotopic anomalies: supernova neighbor or presolar carriers?', *Icarus* **32** (1977), 255.
13. G. M. Spruch, 'New evidence supports supernova origin of Solar System', *Physics Today* **30**, No. 5 (1977), 17.
14. H.-M. Hahn, 'Löste eine Supernova die Entstehung der Sonne aus?', *Bild der Wissenschaft* **14**, No. 8 (1977), 98.
15. J. M. Lattimer, D. N. Schramm and L. Grossman, 'Supernovae, grains and the formation of the Solar System', *Nature* **269** (1977), 116.
16. H. Reeves, 'The origin of the Solar System', *Mercury* **6**, No. 2 (March/April 1977), 7.
17. R. I. Thompson, P. A. Strittmatter, E. F. Erickson, F. C. Witteborn and D. W. Strecker, 'Observation of preplanetary disks around MWC 349 and LkHα 101', *Astrophysical Journal* **218** (1977), 170; *Physics Today* **30**, No. 11 (1977), 19.
18. S. H. Dole, 'Computer simulation of the formation of planetary systems', *Icarus* **13** (1970), 494.
19. R. Breuer, 'Das stärkste Magnetfeld im Kosmos', *Bild der Wissenschaft* **14**, No. 5 (1977), 146.
20. M. H. Hart, 'The evolution of the atmosphere of the Earth', *Icarus* **33** (1978), 23.
21. M. H. Hart, 'Habitable zones about main sequence stars', *Icarus* **37** (1979), 351; see also 'A history of the Earth's atmosphere', *Sky and Telescope* **53**, No. 4 (April 1977), 266.

22. M. J. Newman and R. T. Rood, 'Implications of solar evolution for the Earth's early atmosphere', *Science* **198** (1977), 1035.
23. S. H. Dole, *Habitable Planets for Man*, 2nd edn (American Elsevier: New York, 1970).
24. See Part II: ref. 27.
25. Letter from M. H. Hart, 1 August 1977.
26. J. D. Hays, J. Imbrie and N. J. Shackleton, 'Variations in the Earth's orbit: pacemaker of the Ice Ages', *Science* **194** (1976), 1121; see also *Science News* **110** (1977), 356.
27. F. H. C. Crick and L. E. Orgel, 'Directed panspermia', *Icarus* **19** (1973), 341.
28. S. L. Miller and L. E. Orgel, *The Origins of Life on the Earth* (Prentice-Hall: Englewood Cliffs, NJ, 1974).
29. F. Hoyle and N. C. Wickramasinghe, 'Does epidemic disease come from space?', *New Scientist* **76** (17 November 1977), 402.
30. A. I. Oparin, *The Origin of Life* (Moscow 1924; English transl. Dover: New York, 2nd edn, 1974).
31. M. Eigen, 'Leben' in *Meyers Enzyklopädisches Lexikon* (Mannheim), p. 715.
32. M. Eigen and P. Schuster, 'The hypercycle. A principle of natural self-organization. Part A: Emergence of the hypercycle', *Die Naturwissenschaften* **64** (1977), 541; 'Part B: The abstract hypercycle', **65** (1978), 7; 'Part C: The realistic hypercycle', **65** (1978), 341.
33. T. von Randow, *Die Zeit* (1977).
34. M. Eigen and R. Winkler, *Das Spiel* (Piper-Verlag: Munich, 1975).
35. M. Eigen, 'The origin of biological information', in J. Mehra (ed.) *The Physicist's Conception of Nature* (D. Reidel: Dordrecht, 1973).
36. M. Eigen, 'Self-organization of matter and the evolution of biological macromolecules', *Die Naturwissenschaften* **58** (1971), 465.
37. B.-O. Küppers, 'Towards an experimental analysis of molecular self-organization and precellular Darwinian evolution', *Die Naturwissenschaften* **66** (1979), 228.
38. *Der Spiegel* No. 24 (1977), 213.
39. P. H. Hofschneider, 'Eingriff in die Erbsubstanz', *Universitas* **33**, No. 1 (1978), 21.
40. G. von Boehm, 'Tod nach der 50. Teilung', *Die Zeit* (4 January 1978), 57.
41. A. G. Cairns-Smith, *The Life Puzzle* (Oliver and Boyd: Edinburgh, 1971).
42. *New York Times* (19 April 1977), 29.
43. H. W. Jannasch and C. O. Wirsen, 'Microbial life in the deep sea', *Scientific American* **236**, No. 6 (June 1977), 42.
44. C. R. Woese, 'A comment on methanorganic bacteria and the primitive ecology', *Journal of Molecular Evolution* **9** (1977), 369.
45. G. E. Fox, L. J. Magrum, W. E. Balch, R. S. Wolfe and C. R. Woese, 'Classification of methanorganic bacteria by 16S ribosomal RNA characterization', *Proceedings of the National Academy of Sciences of the USA* **74** (1977), 4537; W. E. Balch, L. J. Magrum, G. E. Fox, R. S. Woese and C. R. Woese, 'An ancient divergence among the bacteria', *Journal of Molecular Evolution* **9** (1977), 305.
46. M. D. Papagiannis, 'Could we be the only advanced technological civilization in our Galaxy?', in H. Noda (ed.) *Proceedings of the Fifth International Conference on the Origin of Life*, Kyoto, Japan, 1977 (Japan Scientific Societies Press: Tokyo, 1978), p. 583.

47. R. W. Kaplan, *Der Ursprung des Lebens* (Georg Thieme Verlag: Stuttgart, 1972).
48. B. M. French, 'What's new on the Moon?', *Sky and Telescope* **53**, No. 3 (March 1977), 164; No. 4 (April 1977), 257.
49. R. Meissner, 'Die Planeten Merkur, Venus und Mars', *Umschau in Wissenschaft und Technik* **77**, No. 10 (1977), 293.
50. J. Eberhart, 'Venus refined', *Science News* **111** (1977), 252.
51. E. Burgess, 'Mars—a water planet?', *New Scientist* **72** (21 October 1976), 152.
52. W. Büdeler, 'Die Marsmonde aus Viking-Sicht', *Bild der Wissenschaft* **14**, No. 3 (1977), 144.
53. Interview with H. Klein: 'Where are we in the search for life on Mars?', *Mercury* **6**, No. 2 (March/April 1977), 2.
54. T. Owen *et al.*, 'Jupiter's rings', *Nature* **281** (1979), 442.
55. R. Smoluchowski, 'The ring systems of Jupiter, Saturn and Uranus', *Nature* **280** (1979), 377.
56. D. M. Hunten, 'The outer planets', *Scientific American* **233**, No. 3 (September 1977), 131.
57. J. L. Elliot, E. Dunham and D. Mink, 'The rings of Uranus', *Nature* **267** (1977), 328.
58. J. C. Bhattacharyya and K. Kuppuswamy, 'A new satellite of Uranus', *Nature* **267** (1977), 331.
59. D. C. Black and G. C. J. Suffolk, 'Concerning the planetary system of Barnard's Star', *Icarus* **19** (1973), 353.
60. H.-M. Hahn, 'Doppelsterne mit bewohnten Planeten', *Bild der Wissenschaft* **14**, No. 12 (1977), 180.
61. E. R. Harrison, 'Has the Sun a companion star?', *Nature* **270** (1977), 324.
62. C. Sagan (ed.), *Communication with Extraterrestrial Intelligence* (MIT Press, Cambridge, Mass. and London, 1973).
63. See ref. 62, p. 361.
64. N. S. Kardashev, 'Transmission of information by extraterrestrial civilizations', *Soviet Astronomy* **8** (1964), 217.

Part II (*pages 91–231*)

1. M. Schmidt, 'A model of the distribution of mass in the galactic system', *Bulletin of the Astronomical Institutes of the Netherlands* **13** (1956), 15.
2. R. N. Bracewell, *The Galactic Club—Intelligent Life in Outer Space* (W. H. Freeman: San Francisco, 1974; Heinemann: London, 1978).
3. G. Cocconi and P. Morrison, 'Searching for interstellar communications', *Nature* **184** (1959), 844; reprinted in ref. 27, p. 160.
4. See Part I: ref. 62, p. 320.
5. S. von Hoerner, 'The search for signals from other civilizations', *Science* **134** (1961), 1839; reprinted in ref. 27, p. 272.
6. N. S. Kardashev, *The Latest Investigations of CETI in the USSR*, Special Publication No. Rp-279 (Institute for Cosmic Research, Academy of Sciences of the USSR: Moscow, 1976).
7. C. Sagan and J. Agel, *The Cosmic Connection—An Extraterrestrial Perspective* (Doubleday: New York, 1973; Hodder & Stoughton: London, 1974), p. 242.

8. W. Sullivan, *We are Not Alone* (McGraw-Hill: New York, 1964; Hodder & Stoughton: London, 1965), p. 328.
9. See Part I: ref. 62, p. 338.
10. See Part I: ref. 62, p. 341.
11. See Part I: ref. 62, p. 336.
12. F. Hoyle and J. Elliot, *A for Andromeda* (Souvenir Press: London, 1962).
13. C. Ponnamperuma and A. G. W. Cameron (eds), *Interstellar Communication: Scientific Perpectives* (Houghton Mifflin: Boston, 1974), p. 80.
14. M. Robertson, 'Computers that learn from their mistakes', *New Scientist* **68** (6 November 1975), 336.
15. See Part I: ref. 62, p. 329.
16. R. Heinlein, *The Moon is a Harsh Mistress* (Putnam: New York, 1966; New English Library, 1969).
17. H. Freudenthal, *Lincos—Design of a Language for Cosmic Intercourse, Part I* (North-Holland: Amsterdam, 1960).
18. See Part I: ref. 62, p. 272.
19. See ref. 7, p. 217.
20. C. E. Shannon and W. Weaver, *The Mathematical Theory of Communication* (University of Illinois Press: Urbana, 1959).
21. S. A. Kaplan (ed.), *Extraterrestrial Civilizations: Problems of Interstellar Communication* (Israel Program for Scientific Translation: Jerusalem, NASA TTF-631, 1971).
22. F. Drake in ref. 13, p. 118.
23. G. Paul, *Unsere Nachbarn im Weltall* (Econ Verlag: Düsseldorf, 1976), p. 107.
24. The staff at the NAIC, 'The Arecibo message of November 1974', *Icarus* **26** (1975), 462.
25. See ref. 23, p. 109.
26. R. S. Dixon, 'A search strategy for finding extraterrestrial radio beacons', *Icarus* **20** (1973), 187.
27. A. G. W. Cameron (ed.), *Interstellar Communication* (Benjamin: New York, 1963).
28. T. B. H. Kuiper and M. Morris, 'Searching for extraterrestrial civilizations', *Science* **196** (1977), 616.
29. Lévy-Leblond, J. M., 'On the conceptual nature of physical constants', *La Rivista del Nuovo Cimento* **7** (1977), 187.
30. F. Drake, 'How can we detect radio transmissions from distant planetary systems?', *Sky and Telescope* **19** (January 1960), 140; reprinted in ref. 27, p. 165.
31. I. S. Shklovskii and C. Sagan, *Intelligent Life in the Universe* (Dell Publ. Co.: New York; Pan Books (Picador): London, 1977).
32. See Part I: ref. 62, p. 260.
33. D. Blake *et al.*, 'Searching for extraterrestrial intelligence; the ultimate exploration', *Mercury* **6**, No. 3 (July/August 1977), 3.
34. A. C. Clarke, *Imperial Earth* (Gollancz: London, 1975), p. 216.
35. B. M. Oliver and J. Billingham (eds), *Project Cyclops: A Design Study of a System for Detecting Extraterrestrial Life* (NASA CR-114445, 1971).
36. B. M. Oliver in ref. 13, p. 141.
37. J. E. Gunn, *The Listeners* (Scribner's Sons: New York, 1972).

38. 'Sowjetisches Riesen-Radioteleskop', *Bild der Wissenschaft* **13**, No. 9, *Akzent* (1976), 2.
39. B. Konowalow, 'Der 6-m-Spiegel im Kaukasus', *Bild der Wissenschaft* **14**, No. 8 (1977), 84.
40. *Science News* **111** (1977), 247.
41. I. Ridpath, 'An ear to the void', *New Scientist* **74** (12 May 1977), 326.
42. See Part I: ref. 62.
43. *Bild der Wissenschaft* **13**, No. 1, *Akzent* (1976), 3.
44. R. N. Schwartz and C. H. Townes, 'Interstellar and interplanetary communication by optical masers', *Nature* **190** (1961), 205; reprinted in ref. 27, p. 223.
45. See Part I: ref. 62.
46. R. Breuer, 'Lebensweg der Doppelsterne', *Bild der Wissenschaft* **13**, No. 9 (1976), 86.
47. R. Breuer, 'Röntgensterne, Neutronensterne und Schwarze Löcher', *Umschau in Wissenschaft und Technik* **76**, No. 12 (1976), 377.
48. See Part I: ref. 62, p. 398.
49. R. Breuer, 'Gravitationswellen erweitern die Astronomie', *Bild der Wissenschaft* **14**, No. 3 (1977), 120.
50. J. M. Pasachoff and M. L. Kutner, 'Neutrinos for interstellar communication', *Cosmic Search* **1**, No. 3 (1979), 2.
51. D. Kirch, 'Tachyonen—Teilchen schneller noch als Licht', *Umschau in Wissenschaft und Technik* **77**, No. 23 (1977), 758.
52. A. S. Lapedes and K. C. Jacobs, 'Tachyons and gravitational Čerenkov radiation', *Nature (Physical Science)* **235** (1972), 6; and references therein.
53. D. Kirch, 'Some theoretical and experimental aspects of the tachyon problem', *International Journal of Theoretical Physics* **13** (1975), 153.
54. H. Yokoo and T. Oshima, 'Is bacteriophage ΦX-174 DNA a message from an extraterrestrial intelligence?', *Icarus* **38** (1979), 148.
55. A. C. Fabian, D. Maccagni, M. J. Rees and W. R. Stoeger, 'The nucleus of Centaurus A', *Nature* **260** (1976), 683.
56. R. Breuer, 'Auf der Suche nach Leben im All', *Bild der Wissenschaft* **17**, No. 1 (1980), 102.
57. K. I. Kellermann, D. B. Shaffer, B. G. Clark and B. J. Geldzahler, 'The small radio source at the galactic center', *Astrophysical Journal (Letters)* **214** (1977), L61.
58. M. Schmidt, 'Quasars', in A. Sandage *et al.* (eds), *Galaxies and the Universe* (University of Chicago Press: Chicago and London, 1975), p. 283.
59. D. O. Richstone and J. Oke, 'The nebulosity near the quasar 3C 249.1', *Astrophysical Journal* **213** (1977), 8.
60. H. Bond, R. Kron and H. Spinard, 'GQ Comae and V396 Herculis: two low-redshift, optically variable QSOs', *Astrophysical Journal* **213** (1977), 1.
61. S. Tsuruta, 'Black hole models for quasarlike objects', *Astronomy and Astrophysics* **61** (1977), 647.
62. L. M. Ozernoy and V. V. Usov, 'Regular optical variability of quasars and nuclei of galaxies as a clue to the nature of their activity', *Astronomy and Astrophysics* **56** (1977), 163.
63. M. J. Rees, 'Accretion and the Quasar phenomenon', in O. Ulfbeck (ed.), *Quasars and Active Nuclei of Galaxies*. Proceedings of a symposium held in Copenhagen, 27 June–2 July 1977. *Physica Scripta* **17** (1978), 193.

64. R. D. Blandford, C. F. McKee and M. J. Rees, 'Super-luminal expansion in extragalactic radio sources', *Nature* **267** (1977), 211.

65. S. Lem, *The Invincible* (Warsaw, 1964; Engl. transl. Sidgwick & Jackson: London, 1973): *Solaris* (Warsaw, 1961; Engl. transl. Walker: New York, 1970; Faber: London, 1971); *Eden* (Warsaw, 1959).

66. A. C. Clarke, *Rendezvous with Rama* (Gollancz: London, 1973).

67. I. Asimov, *Foundation; Foundation and Empire; Second Foundation* (Gnome Press: New York, 1951–3; first published in the magazine *Astounding Science Fiction*, 1942–50).

68. F. J. Dyson, 'Search for artificial stellar sources of infrared radiation', *Science* **131** (1960), 1667; reprinted in ref. 27, p. 111.

69. F. J. Dyson, 'The search for extraterrestrial technology', in Marshak, R. E. (ed.), *Perspectives in Modern Physics* (Interscience: New York, London and Sydney, 1966), p. 641.

70. S. von Hoerner, 'Population explosion and interstellar expansion', in E. Scheibe and G. Süssman (eds), *C. F. von Weizsäcker Festschrift* (Vandenhoeck & Rupprecht: Göttingen, 1973), 221.

71. N. H. Kinard, R. L. O'Neal, J. M. Alvarez and D. H. Humes, 'Interplanetary and near-Jupiter meteroid environments: preliminary results from the meteroid detection experiment', *Science* **183** (1977), 321.

72. See Part I: ref. 62, p. 371 (Appendix D).

73. See Part I: ref. 62, p. 390.

74. P. Dyal, *Symposium on Recent Results in Infrared Astrophysics* (NASA TM X-73, 1977), p. 190.

75. F. J. Dyson, 'Intelligence in the Universe', *Mercury* **1**, No. 6 (November/December, 1972), 9.

76. H. Zimmer, 'Himmelsdurchmusterung im Infrarot', *Bild der Wissenschaft* **15**, No. 1 (1978), 106.

77. See Part I: ref. 46.

78. M. H. Hart, 'An explanation for the absence of extraterrestrials on Earth', *Quarterly Journal of the Royal Astronomical Society* **16** (1975), 128.

79. C. Ponnamperuma, *The Origins of Life* (Thames & Hudson: London, 1972).

80. See ref. 13, p. 45.

81. J. A. Ball, 'The zoo hypothesis', *Icarus* **19** (1973), 347.

82. E. M. Jones, 'Colonization of the Galaxy', *Icarus* **28** (1976), 421.

83. I. Asimov, 'Homo Sol', in *The Early Asimov*, Vol. 2 (Panther Books: St Albans, Herts, 1974), p. 13.

84. C. Sagan, 'On the detectivity of advanced galactic civilizations', *Icarus* **19** (1973), 350.

85. I. Ridpath, 'The mini-planet', *New Scientist* **76** (17 November 1977), 406.

86. See Part I: ref. 62, p. 337.

87. E. Fasan, *Relations with Alien Intelligences* (Berlin Verlag: Berlin, 1970), p. 71.

88. F. J. Tipler, 'Extraterrestrial intelligent beings do not exist', *Quarterly Journal of the Royal Astronomical Society* **21** (1980), 267.

89. F. Hoyle and N. C. Wickramasinghe, 'Does epidemic disease come from space?', *New Scientist* **76** (17 November 1977), 402.

90. B. Küppers, 'Wissenschaftstheoretische Probleme der Biologie/Evolution/Zufall', in

Speck, J. (ed.), *Handbuch wissenschaftstheoretischer Begriffe* (Vandenhoeck & Rupprecht: Göttingen, 1978).

91. L. E. Navia, *Das Abenteuer Universum* (Econ Verlag: Düsseldorf, 1977).
92. J. Hermann, *Astrobiologie* (Franckh'sche Verlagshandlung: Stuttgart, 1974), p. 132.
93. B. M. Oliver, 'Proximity of galactic civilizations', *Icarus* **25** (1975), 360.

Part III (*pages 233–276*)

1. J. K. Beatty, 'Voyaging to the outer planets', *Sky and Telescope* **54**, No. 2 (August 1977), 95.
2. *Der Spiegel* No. 54 (1977), 132.
3. See Part II: ref. 2.
4. See Part II: ref. 13.
5. See Part II: ref. 3.
6. C. Sagan, 'Direct contact among galactic civilizations by relativistic interstellar spaceflight', *Planetary & Space Science* **11** (1963), 485.
7. S. von Hoerner, 'The general limits of space travel', *Science* **137** (1962), 18; see also Part II: ref. 70.
8. See Part II: ref. 27, p. 121.
9. See Part II: ref. 27.
10. 'Project Daedalus: an interim report on the BIS starship study', *Spaceflight* **16** (1974), 356.
11. R. W. Bussard, 'Galactic matter and interstellar flight', *Astronautica Acta* **6** (1960), 179.
12. See Part II: ref. 13, p. 172.
13. C. Powell, 'The effect of subsystem inefficiencies upon the performance of the ram-augmented interstellar rocket', *Journal of the British Interplanetary Society* **29** (1976), 786.
14. W. B. Roberts, 'The relativistic dynamics of a sub-light-speed interstellar ramjet probe', *Journal of the British Interplanetary Society* **29** (1976), 795; and references therein.
15. G. Marx, 'Interstellar vehicle propelled by terrestrial laser beam', *Nature* **211** (1966), 22; see also Part I: ref. 62.
16. A. Kantrowitz, 'Laser propulsion to Earth orbit' (preprint, Avco-Everett Research Laboratory, 1976).
17. T. A. Heppenheimer, *Colonies in Space* (Stackpole Books: Harrisburg, Pa., 1977), p. 143.
18. L. A. Booth, D. A. Freiwald, T. G. Frank and F. T. Finch, 'Prospects of generating power with laser-driven fusion', *Proceedings of the Institute of Electrical and Electronic Engineers* **64** (1976), 1460.
19. See Part II: ref. 7, p. 247.
20. See Part II; ref. 82.
21. See Part II; ref. 78.
22. See Part II: ref. 70.
23. G. K. O'Neill, 'A Lagrangian community', *Nature* **250** (1974), 636; *The High Frontier: Human Colonies in Space* (W. Morrow & Co.: New York, 1977); *Mercury* **3**

No. 4 (July/August 1974), 4; *Bild der Wissenschaft* **13**, No. 5, *Akzent* (1976), 5; *Penthouse* (August 1976), 86.

24. G. K. O'Neill, 'The colonization of space', *Physics Today* **27**, No. 9 (1974), 32.

25. *Journal of the British Interplanetary Society* **29** (1976): articles by P. J. Parker (p. 764), G. L. Matloff (p. 775), I. R. Richards and P. J. Parker (p. 769); **30** (1977): articles by A. V. Cleaver (p. 283) and M. M. Hopkins (p. 289).

26. E. U. Condon, *Scientific Study of Unidentified Flying Objects* (Bantam Books: New York, 1968; Vision Books: London, 1970).

27. C. Sagan and T. Page (eds), *UFOs—A Scientific Debate* (Cornell University Press: Ithaca, NJ and London, 1972).

28. P. J. Klass, *UFOs Explained* (Random House: New York, 1974).

29. A. Schneider, *Besucher aus dem All* (H. Bauer Verlag: Freiburg, 1976).

30. J. A. Hynek, *The UFO Experience: A Scientific Inquiry* (H. Regnery: Chicago, 1972; Abelard-Shuman: London, 1972).

31. P. A. Sturrock, review of *UFOs—A Scientific Debate* in *Science* **180** (1973) 593.

32. H. Hartung, 'Bauwerke der Mayas weisen zur Venus', *Umschau in Wissenschaft und Technik* **76**, No. 16 (1976), 526.

33. R. K. G. Temple, *The Sirius Mystery* (Sidgwick & Jackson: London, 1976).

34. K. Herbert, *Warum?* (March 1977), 10.

35. E. von Däniken, *According to the Evidence* (Souvenir Press: London, 1977).

Index

QB54 .B69813 1982   010101 000
Breuer, Reinhard A., 1946
Contact with the stars : the

0 2002 0018178 0

YORK COLLEGE OF PENNSYLVANIA 17403